SUNSHINE

Also by Robert Mighall

A Geography of Victorian Gothic Fiction:
Mapping History's Nightmares

SUNSHINE

One Man's Search for Happiness

ROBERT MIGHALL

JOHN MURRAY

First published in Great Britain in 2008 by John Murray (Publishers)
An Hachette Livre UK company

1

A CIP catalogue record for this title is available from the British Library

Hardback ISBN 978-0-7195-9510-3
Trade paperback ISBN 978-0-7195-9547-9

Typeset in Fairfield LH Light by Servis Filmsetting Ltd, Manchester

Printed and bound by Clays Ltd, St Ives plc

John Murray policy is to use papers that are natural, renewable
and recyclable products and made from wood grown in
sustainable forests. The logging and manufacturing processes are
expected to conform to the environmental regulations of the
country of origin.

John Murray (Publishers)
338 Euston Road
London NW1 3BH

www.johnmurray.co.uk

To lost love

CONTENTS

'The Sun is God.'

J. M. W. Turner's deathbed declaration

Late September a few years ago, at the end of a truly lousy summer, the sun finally appeared after weeks and weeks of gloom. I went for a walk that afternoon, and passed by an old people's home. The French windows had been thrown open, and most of the residents wheeled out into the garden to enjoy this belated appearance of the sun. They seemed to cling to it like life itself. For some, this might well be the last time they would feel its caress. I moved on; imagining, but unable to witness, the moment when the shadows returned, and they were wheeled back inside to the mocking surrogate of the flickering TV screen.

A few scattered rays of sunshine in a dull September, years ago: barely a blip on the statistical record for that month, but for me the moment was epiphanic. Deeply touched by what I had seen, I felt a need to understand why the sun matters to us so. Why it matters so much to me.

1

PRODIGAL SUN

> I never saw a man who looked
>   With such a wistful eye
> Upon that little tent of blue
>   Which prisoners call the sky,
> And at every drifting cloud that went
>   With sails of silver by.
> Oscar Wilde, *The Ballad of Reading Gaol* (1898)

I am, and have been for as long as I can remember, obsessed with sunshine. Maybe even addicted. Like some heliotropic plant, my whole mood and outlook is profoundly affected by the fickle play of the photosphere. There used to be a game on *Playaway*, the children's TV show, which would start with the presenter (usually Johnny Ball) grinning insanely at the camera. He would then pass his hand slowly in front of his face from below, to reveal a countenance changed to near tearfulness; this returned to manic glee when he reversed the exercise.

Clouds have the same effect on me. Sun out, smile; sun in, frown. Especially in the summer, when I nurture some mythical idea of what the sun should be doing, and anxiously scrutinize the skies in the hope that it might just clear. Or if the weather's fine, I look for the inevitable signs that the skies will revert to their default setting. With obsession comes superstition, and an almost childish inability to temper devotion with reason.

For me the sun and clouds are mortal enemies, mutually exclusive, and it is still hard for my poor unscientific mind to accept that the sun is actually responsible for clouds. I'm told its rays lift moisture from the earth and the seas, thus making the clouds that can bring rain. This is what meteor-ologists call the 'weather machine', and it is of course essen-tial for maintaining life on this planet. But that basic fact still seems like a terrible betrayal to me. And so each summer is like a battle fought out above my head: between my side, the sunny side, and the battleship grey battalions laying siege to my world.

My neighbours call me 'Gecko Boy', from my habit of climbing a ladder against our building to catch the last few rays of the descending sun. I've sunbathed before work, at 7.30 in the morning; and know where and for how long there will be patches of sunlight after work in the summer, and reli-giously do my rounds of them in strict order on my way home. I only walk on the sunny side of the street, and will go out of my way, or be late for appointments, to avoid shade. I've per-plexed waiters the world over by choosing the only table out in the sun on their terraces, and refusing their well-intentioned attempts to erect parasols over my poor crazed English head.

If the sun's out and I'm not, I'm like some junkie climb-ing the walls. And so I take furtive sun breaks as others take fag breaks during the working day. Why on earth not? Surely I'm as entitled to a ten-minute fix of my favoured carcinogen as they are. For each man dies by the thing he loves . . . And as for love, I've always been attracted to dark-haired, dark-skinned women in some unconscious attempt at consummation with the south, with a flesh and blood avatar of my distant mistress – my 'true mistress', as one former girlfriend tetchily remarked. My adoration has caused its fair share of conjugal conflict over the years. I've yet to find a girlfriend who has shared this love or been prepared to share mine. Or who has accepted why I need to be outside at 7.00 a.m. on a sunny Saturday, instead of – as John Donne nearly put it – making our own sun within.

Yes, I'm an obsessive, but surely only an extreme version of a national or even hemispherical type. The English are reputedly obsessed by the weather, and as northerners we have an unquenchable sense of solar deficit. You need only look around you. If the sky is blue and the sun is shining, people are more likely to be smiling, relaxed, at one with and more open to the world. If it's raining, cold and overcast, they will huddle into themselves, their faces as gloomy as the soggy wool coverlet under which we cower for so much of our lives. You get a sense that if the sun appeared more often this land might be a little less green, but a good deal more pleasant. We'd bottle sunshine if we could.

And that is exactly what I've tried to do. To capture and celebrate something to which most of us feel a potent primal attraction, but which has been strangely denied serious or sustained attention in print. When compared with other sensual stimulants such as sex or wine it is very poorly served. Yet, sunshine has its champagnes, its plonks and its vintages. It can be young, full-bodied, boast tremendous finish or, as too often happens, be corked by fast-gathering clouds. There are sunnings you glug, those you savour and those glorious

5

vintages (76, 95, 03, 06) you lay down in the cellars of memory. The joys of sunlit sensation can be fondly annotated by the besotted sun-bibber. The pellucid purity of early spring's young sunning (Wedgwood blue with promising hints of riper Matisse). The mellow fruitfulness of a late-bottled Indian vintage. The peppery prickling of the fully-rounded day-long scorcher, blushed with generous tannins. The soft buttery tones of a rare summer day's enchantment, with a final burst of marmalade for a crepuscular crescendo in the west.

If you can sing a rainbow, then perhaps you can write some sunshine. Or at least try to pin it down in words, tune into some of its wavelengths and reveal some of its mysteries. I'm not principally talking about the Sun as a personified deity; but about the stuff that hits you with a sense of homecoming when you step on to the tarmac of a foreign airport, or that the cat – who knows nothing of myth or metaphysics – serenely worships. That stuff that thrills the eyes with chromatic pleni-tude, penetrates the soul with its soothing or energizing caress, and lifts you into its heady kinetic rush when it bursts through the clouds or streams into a room.

Ah, so it's one of those books, a friend said: a bit like *Thing: The* [foodstuff / colour / essential stationary item, etc.] *That Changed the Face of History*. But really it's the opposite of these books. Whilst they take a supposedly insignificant object and explain its enormous importance in the grand nar-rative of history, I'm attempting the reverse.

The sun and what it affords is universally acknowledged as quite important. It's the source of nearly all life on earth for a start. And it's truly marvellous to consider what it does. Plants converting sunlight into sugar; animals eating those plants to gain their own energy; humans eating those plants or animals to provide the energy needed to build their cities, wage their wars, write their books, and make more humans; humans being eaten by worms just to make things interesting and put us in our place – our place in the great chain of being, forged and maintained by that vast furnace in the sky.

It's pretty impressive. But rather too impressive for most of us to remember too often, or contemplate for too long. Because we can't step outside it, it's difficult to comprehend – like chasing our shadows or attempting to see eternity in a grain of sand. The sun might be considered too big and old and omnipresent to have a human history and be reduced to an everyday scale. But that's what I'm attempting.

It's no doubt a great relief to Galileo's shade that he recanted his heretical views about the heliocentric universe. Had he or Copernicus been martyred for their belief that the earth rotates around the sun rather than vice versa, their deaths would have been largely in vain. For most of us live in a universe profoundly untouched by most of the break-throughs in solar science. Our sun rises and falls for us, and moves across our skies. We have to see it to believe it's there, and we can lament its apparent absence as if it really had forsaken the universe. It can be eclipsed simply by some thoughtless individual standing up, stretching and deciding they will go in search of an ice cream. Ours is a sun we can get out 'into', get 'some of', or secure a place 'in'. We can spend a fortune following it around the globe, or a whole day following it around the garden.

This is the sun I am pursuing. The one that revolves around us, happens for us, and whose absence or presence has a profound effect on the quality of our lives day to day. This is more a work of poetics than physics.

A book about sunshine is also in danger of being a book about everything: everything it shines on, brings to life and that reflects it. I've had to cut a path through this ubiquity, and to focus on those aspects of sunshine that appear particularly relevant to us here and now, in the world we inhabit today. I have therefore not attempted to tackle in any systematic way the myriad mythological, religious or spiritual meanings of the Sun, except when they appear to have relevance or resonance for us still. By us, I mean the northern hemisphere, and principally this little wind-fretted, rain-soaked island on the extreme edge of north-western Europe.

Which brings me to the rather delicate issue of Englishness vs. Britishness. In general, I'm talking about the British Isles, the weather we get here, and the behaviour and attitudes this has encouraged. Yet, at times I do specifically talk about Englishness, and use Kate Fox's diagnosis of what she calls English 'social dis-ease', in her *Watching the English*, as a useful authority on what this is supposed to entail. I'm mostly writing about what I know, and about how the skies look above my own bunker. As those skies happen to brood low over London, I might appear guilty of a certain parochialism at times. Apologies in advance.

This is an obsessive's quest to understand the forces that have shaped his outlook and identity. Some of these may be universal – physical rays affecting neurological synapses – others are probably a product of history, cultural conditioning and the complex coaxings of desire. Some will reflect a national outlook, and express a collective response to life lived on the edge of a continent and menaced by capricious forces. And some I must be held entirely responsible for. So, sometimes when I say 'we', I should probably say 'I'. A degree of solipsism is inevitable when we're dealing with an obsession; sol-ipsism when dealing with this one.

There is a story, possibly apocryphal, about the last days of

8

the painter J. M. W. Turner. Turner, like his contemporary Constable, was obsessed with the weather, and was especially fascinated with capturing or, we might even say, dramatizing sunlight. John Ruskin claimed that Turner was 'a Sun-worshipper of the old breed'. This devotion betrays itself in countless paintings where the ostensible subject of the painting – a ship, a grand building or landscape, an historical or mythological subject – is upstaged by a stupefying solar spectacle. According to one art historian, '*The Sun rising through vapour*, the title of a painting he exhibited in 1807, could easily be the name of half his work.'[1] This obsession appears to have stayed with him to his very last days. According to Ruskin, the artist declared that '"The Sun is God", a few weeks before he died with its setting rays on his face.'[2]

Ruskin is referring to and embellishing an anecdote apparently recounted by Turner's landlady and companion, Mrs Booth, shortly after the artist's death. Turner had moved into a cottage in Chelsea, so he could be close to the river he loved. This house had a platform, from which he would watch the river for hours.

> During his last illness the weather was dull and cloudy, and he often said in a restless way, 'I should like to see the sun again.' Just before his death he was found prostrate on the floor, having tried to creep to the window, but in his feeble state he had fallen in the attempt. It was pleasing to be told that at the last the sun broke through the cloudy curtain which so long had obscured its splendour, and filled the chamber of death with a glory of light.[3]

I don't care if it's not true. It should be, and it is significant that this legend has emerged and endured. For me it is no less tragic and moving than Lear's wish to see his daughter Cordelia again before he dies. The absence, the pathetic yearning, the rapturous communion when the sun belatedly appears, and the final declaration of divinity. The same conditions and sentiments shape the pages that follow. The same declaration stands as their epigraph.

If the Sun is God, then this is my version of his theology.

2

NORTHERN SKY

'Yes, of course, if it's fine tomorrow,' said Mrs Ramsay
. . . 'But,' said his father, stopping in front of the
drawing-room window, 'it won't be fine.'
Virginia Woolf, *To the Lighthouse* (1927)

'Weather permitting' – is there a more English expression? A politely genteel version of the Muslim's *inshallah* (God willing), invoked for barbecues, village fêtes, picnics and sporting events year after year. Its recurrence makes me wonder if we've completely escaped the thraldom of the ancient weather gods. I know I haven't. Personally, I'd be quite happy to paint myself blue, dress my hair in twigs, and dance stark naked in a wood at full moon each November if it meant the next summer would improve on the UK standard of three days of sun followed by a thunderstorm.

Scarcity and capriciousness have always defined our relationship with the sun, something that goes back deep into the mists, snows and rains of time. To understand our love of sunshine, we need to penetrate those mists, to see how much of what we do and feel today was shaped by such ancestral legacies. Here, then, is my rather partial prehistory of our obsession with the sun. Or, if the truth be told, a good old moan about the British weather, combined with a few historical speculations.

Of course nearly all cultures on earth have worshipped the sun at some point in their history. As Jacquetta Hawkes puts it, the sun was 'the beginning of all the Fathers which are in Heaven'. This is hardly surprising. Early humanity worked out what we have to remind ourselves of with an effort – we *are* because the sun *is*. We may no longer have much need of gods, but we still need the sun. They needed both. With consciousness came a host of questions and demands, and you can understand why that great life-bestowing, all-seeing, time-regularizing and truly awe-inspiring phenomenon was a fair candidate for the job.

From the long and illustrious pageant of ancient sun-worshippers, I will ask my personal favourites to step forward: Akhenaten, the Egyptian pharaoh of the eighteenth dynasty, and Diogenes the Cynic (399–323 BC).

The sun was worshipped in many forms and aspects in ancient Egypt. When Amenophis IV became pharaoh, he pledged his exclusive allegiance to the Aten, 'the bright Disc of the day', and with this established the world's first monotheistic religion. The young pharaoh decided that the physical sun shining in the sky was the one true god. He changed his name to Akhenaten (meaning the Aten is satisfied), and built a new city, Akhetaten (the Horizon of the Aten), where he worshipped this god, his father.

He no doubt believed fervently in the divinity of the sun, but he also appears to have been an enthusiastic sunbather. He spoke of how 'the Aten shone on him in life and length of days, invigorating his body each day', betraying an appreciation of the physical, sensual and even aesthetic radiance it bestowed. His monotheism may have derived from a mono-mania with which I can certainly identify:

> Thou art shining, beautiful and strong;
> Thy love is great and mighty,
> Thy rays are cast into every face,
> Thy glowing hue brings life to hearts.[1]

The rays of the sun were what our young king appreciated and worshipped, and that's why he deserves special mention.

The first monotheist has the additional claim to fame of being one of the few figures from history to have a pair of pants named after him. And skimpy ones at that. This distinction was bestowed upon him by some of the pioneer nudists of the early twentieth century. As the next chapter will show, the nudists were the first modern sun-worshippers to make a cult of the physical experience of sunshine. According to the *Sun-Dial*, a publication circulated at one of the first nudist camps in Britain, 'Akhenaten Slips' were 'practically translucent to ultra-violet rays', and allowed near nudity to those who weren't quite prepared to strip off entirely.[2]

Diogenes the Cynic was also canonized by the nudists. One of the oldest surviving UK 'sun clubs' was named after him. This distinction derives from something Diogenes supposedly said to Alexander the Great. The king was an admirer of the irreverent philosopher, and visited him to ask if he could do anything for him. 'Get out of my sunshine', is the apocryphal reply. Again, this is not the worship of the sun as a god (Helios or Apollo), but a simple appreciation of sunshine as a physical experience. This is sun-worship as we recognize it today. In antiquity, however, this term principally meant exactly that, the veneration of the star as a personified and all-powerful being.

If you've got to worship something, then the thing that keeps you warm, makes your food grow, and gives you a nice tan is pretty worthy of veneration. It's a wonder we ever needed to upgrade to something more complicated. The principal attraction of the sun as a deity must be the fact that you can actually see it. To the primitive mind, myself included, a god you can see, that makes its benefits obvious and doesn't move in mysterious ways (keeping a sure and steady course across the heavens, with only the occasional apocalyptic eclipse to keep devotees on their toes), makes perfect sense, and deserves due respect.

But in time, this predictability probably became something of a drawback. Humankind began asking for a bit more sophistication from its deities. The unseen is a more potent source of awe and fear than the manifest, and in due course took over. The Hebrew God appears to forsake His chosen people, and plays hard to get; only manifests Himself in burning bushes or through commandments, or some nasty retributions – floods, pestilence, and so on. His Son turns up, then off he goes again, promising to return. Have faith. Let the spirit move you. Mystery becomes the order of the day.

If that's the prerequisite for the more discerning devotee, then the Sun, boringly regular, predictable and clearly manifest, was destined to be dethroned. And so it happened. In 325 the Emperor Constantine, who had been a fervent sun-worshipper, changed his allegiance, and pronounced Christianity the official state religion of the Roman Empire. The Sun is replaced by the Son; his symbolic assets were stripped and absorbed into the ascendant monotheistic regime. The festivals celebrating the returning, exultant or retreating sun were rebranded to ensure buy-in from the pagan cultists (most notably 25 December, the festival of the sun's birth). When depicting their new lord, they gave Him a golden halo like the one shining from the solar disc He had replaced.[3] In those days it was Jesus who was wanted for a sunbeam, and His iconographers were more than happy to oblige.

Most of these goings-on took place in or around the cradles of civilization: where the sun shines. In the north we've always had a rather different relationship with this deity. Whilst the predictability of the Mediterranean sun rendered it obsolete for mature veneration, the lack of predictability has made it an enduring object of desire in the north.

Our ancestors also worshipped the sun, as attested by the numerous monuments scattered throughout the British Isles. In Loughcrew in the Irish midlands there is a tomb complex dating from around 3300 BC. One tomb in particular displays what David Whitehouse claims are 'some of the most beautiful examples of Neolithic art in Ireland'. These were designed exclusively to be illuminated by the passage of the sun entering the chamber at the spring and autumn equinoxes. This must have been an awe-inspiring spectacle, and have had major significance. As Whitehouse wonders:

Why did those people, struggling to survive among the forests and hills of northwest Europe not long after the retreat of the ice, build such things? They were a great investment in time and energy . . . Clearly, their connection with the Sun was important and was something beyond their day-to-day cares, so they carved and heaved stone, watched and wondered and they left us symbols whose meanings we can only guess.[4]

And then of course there is Stonehenge – one of the most famous, impressive and sophisticated Neolithic structures in the world, whose solar alignment implies that it served a profound ritualistic and spiritual function. As such, it was probably the most redundant public edifice to be built that side of the Millennium Dome. One can't help being stricken with pity at the idea of our forebears (in the days before easyJet) being dedicated sun-worshippers. All that effort, all that precision, all that anticipation. Taking centuries to construct an edifice whose *raison d'être* is a moment when the sun is in a particular place and does a particular thing. And then it's overcast (as it was the one time I attended the summer solstice). It puts the occasional rained-off picnic in perspective.

The tribes gather for the solstice:
sunset, 20 June 2007. Sunrise? It didn't.
The skies were as grey as the stones

Imagine the scene at Loughcrew. You've geared the people up for something spectacular. The exquisite carvings are calculated to a nicety, and ready to be ignited into awe-inspiring life by your solar pyrotechnics. You've brought it in on time and on budget (give or take a century and a noggin of woad); your status as a priest or wise man depends on this performance, and the god lets you down. Two or three of these and you might start thinking about a change of vocation, or watching your back. This could go either way: it could lessen the appeal of such a damp squib of a religion – what's the point of worshipping something that fails to turn up every time we're promised something amazing? Or it could make that god even more important, awesome and revered.

If you're truly a wise man, you might apply the theological version of *The Rules*, the bible of successful dating. I'm told this stipulates that you should ideally wait about three days after a date before you phone the other person. By then, you've worked on the psychology of desire, by making the other person anxious and valuing you more. Three years failing to show, and then wow! On the fourth year he rose again, and there was much rejoicing. There's impressive for you, there's mysterious. And there is love. Such love. The start of a very long, very devoted affair. We've been besotted ever since.

As I write these words, I stare out of my window at a uniformly grey and brooding sky. A few lacklustre spots tinkle against the panes. It's the middle of August, and a Sunday. It's been like this for about three weeks. I close the blinds, shut out the reality and go online. As I suspected, the weather for my favourite resort in the south-west of Spain displays five blazing sun symbols for the coming week. Jackpot. I picture to myself the beach I know so well. It's 2.00 p.m. local time, and the fortunate locals will have enjoyed a few leisurely hours on the beach. It's quite a pious city, so they probably went to Mass early in the day, performing a ritual to an unseen God whose name they gave to the day. *Domingo* – the Lord's day.

By now they will be gathering up their things, dismantling their parasols (for they employ such devices here) and heading off home for a two-hour lunch followed by a siesta. Thus demonstrating their implicit faith in a different god – the one very much in evidence and making his presence felt by beating relentlessly down from on high. They have faith that he will still be there in a kinder guise in a few hours' time, when they can emerge from their shady repose and enjoy a few hours of his softer beneficence. Before saluting his lingering descent into the sea, having a gossip and a stroll, and then thinking about their stomachs again.

Not everyone on the beach this day will have observed this ritual or displayed this faith. The northern Europeans will have remained vigilantly at the posts they probably marked out as early as 9.00 this morning, locked in fervent, desperate adoration of a god they can both see and feel: a god they will not let out of their sights all day, perhaps occasionally scanning the skies for fleecy heretics who might present a challenge to their creed.

They call this day Sunday, *Söndag* or *Sonntag* (retaining the Romans' *dies solis*, which the predominantly Catholic southern nations were happy to rebrand as the Lord's Day). Protestantism is marked by its tendency to challenge, by its need to witness and worry over what other sects are prepared to take on trust. As Oscar Wilde observed, the Church of

England 'is the only Church where the sceptic stands at the altar, and where St Thomas is regarded as the ideal apostle'.[5] Might there not be a climatic connection behind this attitude? Show me Thy radiant face, O Lord, so I might believe – a naive hope that if we honour the sun with words and symbols he might remain in reality.

In the UK Trademark Registry there are 497 registered trademarks using the sun as a symbol. There are 353 for the cross, 366 for the crown, and only 239 for the flag. In this Christian, monarchical, flag-waving country, the sun remains the most potent symbol of devotion.

There appears to be something lodged deep in the collective psyche that makes us pledge our anxious devotion to this remote indifferent star. Could it be some tribal memory of the last great Ice Age? According to climatologist Philip Eden, the most recent glacial period was 'at its most intense from 24,000 to 14,000 years ago', when most of northern Europe above 45 degrees latitude was covered in permafrost and decidedly chilly. Eden explains how in roughly 8500 BC, 'the last glacial phase ended and the present interglacial phase began'.[6] So, technically we're in an 'interglacial period', and should enjoy it while it lasts. Yet, I can't help thinking we haven't quite got over the last withdrawal of the sun, and are still traumatized by the recollection.

The anxiety has left its mark on the mythology and mindset of the northern peoples. The Twilight of the Gods is a potent idea that resonates through Nordic legends, and the madness that grips northern Scandinavia when the sun returns each year must express more than one winter's worth of relief. Such legacies have had a profound effect on our outlook, and left their stamp on our culture – especially in the UK.

GREY SKY THINKING

In the far north they don't expect to see much of the sun each winter, so they hunker down, crack open the vodka, buy four months' worth of sunbed coupons, and get on with it. Whilst the UK's solar deprivation is not so extreme, our attitude to the weather appears to be. We're famed for it. This is not really surprising: as a recent book about the British weather declares, the British Isles is 'the most weather-affected place on earth'.[7] Affected in all senses. We are tormented by the caprices of four competing air masses, and we allow this uncertainty to get to us. Uncertainty is the key to understanding our obsession, and perhaps much of our outlook and behaviour.

In most Romance languages time and weather are the same word. This suggests a certain regularity in the seasons, a way of measuring time until clocks were invented. Sundials monumentalize this relationship. Our word 'weather' even sounds like a question. Although there is no etymological connection, this is somehow fitting (indeed, the *Oxford English Dictionary* suggests 'weather' and 'whether' were interchangeable spellings until the sixteenth century). At the very least, this separation suggests a different relationship with the heavens, perhaps even with time itself. As Samuel Johnson famously observed back in 1758: 'an Englishman's notice of the weather, is the natural consequence of changeable skies, and uncertain seasons.'[8] The extraordinary advances in meteorological science since then have done nothing to stem this flood of small talk.

But what does it signify? According to Kate Fox, very little. Her *Watching the English* dismisses our tendency to talk about the weather as mere 'social grooming', the human equivalent of apes foraging in each other's fur; a polite liturgy for a rule-bound, terminally awkward culture, but without any genuine interest in the subject itself. I'm not so sure. This ritualistic explanation doesn't explain the extent to which weather crops up in our lives: from the metaphors we use to the art we create, or the column inches devoted to the subject in the daily press. Nor does it explain why the longest-standing meteorological records derive from these islands (the first weather diary – yes, a diary obsessing over the weather – dates from 1337), sustained by the extraordinary number of weather chat-rooms based in the UK.[9]

Weather chat-rooms are the preserve of the more extreme weather-watchers, and it's enlightening to eavesdrop on the avalanche of speculation that ensues in them after the merest rumour of snow. The general public is, however, a good deal more partial in its interest, and might be called fair-weather friends to meteorology. As Kate Fox observes, there is an unofficial weather hierarchy that operates in this country. Sunny and warm is at the top, followed by sunny and cold, and rainy and cold least popular of all. (Snow is an honourable exception, but for Christmas, not for life.) This suggests that most of us are not interested (or even uninterested, according to Fox) in the weather *per se*, but betray a distinct preference for a certain type of weather. We do like sun, we don't like clouds and rain (cloud-spotters excepted – we're also a nation of eccentrics). And surely it's only a hard core of anoraks who wouldn't forego a fascinating subject of conversation for a little more certainty, and a lot more sunshine.

Which is a shame, as Fox neglects to point out the obvious inverse relation between what we like and what we generally get, or believe we get. So I conducted my own research to complete this picture. I asked a sample of one hundred people what their favourite weather was. The results clearly confirmed Fox's findings, with sunny and hot or sunny and mild coming out tops. I then asked respondents to indicate the weather we generally get in the UK. Whilst 97 per cent of the sample preferred sun, 89 per cent thought the average weather in the UK was cloudy. (And this was conducted in August 2006, after a particularly good summer.) There would appear to be some pretty disgruntled people out there.

Whether this is a true picture or not, it is the general perception of a sample comprising mostly UK residents and some tourists. Cloudy and damp is how we see our weather, and how others see it. This view was supported anecdotally by most people with whom I discussed the UK weather on my travels. The London cabbie, infallible oracle of popular sentiment, never once departed from this perception (even if the sun was shining brightly outside, he'd have a moan about the weather). An Italian neighbour of mine, whom I encountered sunbathing, told me that when she first came to England she couldn't understand the fuss everyone made about the sun coming out. Now she does. Eight years on and she's making the most of it with the rest of us. It's an article of faith with us, and part of our national identity. This clash between preferences and perceptions must have had some effect on our collective psyche.

In Romance languages, the word for heaven and the sky is the same. In French it is *ciel* (also a shade of blue) and *cielo* in both Italian and Spanish, deriving from the Latin *caelum*: sky or heaven. In the south they can see or imagine heaven when they look up and see infinite celestial blue. Our word 'sky' derives from the Old Norse word for cloud. The Anglo-Saxon word *wolcen* also meant both sky and cloud. Hardly a promising start, suggesting that the realm we inhabit is closer to the Stygian gloom of the other place than the bright prospect of heaven. A survey of skies typically depicted by different national schools of art assessed, among other things, the visibility and cloud cover in a large sample of paintings. The British school won hands down for low clouds, overcast skies and poor visibility.[10]

We can spend what seems like eternity under this cloudy incubus, and it's often a genuine surprise to rediscover that skies are available in blue. Living in London, I find this particularly oppressive, with every dimension hedged in with the same monotonous colour – the colour of bad potatoes, school uniforms and our less memorable politicians. A pilot friend of mine told me the best part of his job was his guaranteed glimpse of blue each day. He did turnarounds from London's City airport to Edinburgh and Glasgow, so this didn't derive from his destinations. But a few times each day, he'd be lifted above the grey afflicting those he'd left below. He recommended the pilot's job to anyone suffering from SAD.

The British have long had a reputation for melancholia. It's hardly surprising. Grey has pulled a lid over our lives, seeped into our souls and left a dark, damp stain on our culture. It is part of our unofficial identity. When J.-K. Huysmans' character the Duc des Esseintes imagines London in *A Rebours* (Against Nature; 1884) he anticipates 'an immense, sprawling, rain-drenched metropolis, stinking of soot and hot iron, and wrapped in a perpetual mantle of smoke and fog'.[11]

Fog is perhaps the quintessentially English, or rather, London, condition. It is also the most aesthetic; a rarefied 'atmosphere' that is now largely associated with a specific time and place – late Victorian London. There is some truth in Oscar Wilde's quip that the Impressionists (following Turner) created the London fogs, or at least our ability to see their aesthetic qualities:

> Where, if not from the Impressionists, do we get those wonderful brown fogs that come creeping down our streets, blurring the gas-lamps and changing the houses into monstrous shadows? . . . There may have been fogs for centuries in London. I dare say there were. But no one saw them . . . They did not exist until Art had invented them.

Yes, there had been fogs before – Tacitus mentions them – but they had been a nuisance, a cause for complaint. 'Where the cultured catch an effect,' Wilde continues, 'the uncultured catch cold.'[12] Monet was evidently highly cultured, as he loved the London fogs, and winter was his favourite season to paint London. He and Whistler were perhaps Britain's first known, perhaps only, weather tourists.

The 'London particular' (as Dickens refers to it) became the capital's signature effect in the nineteenth century, and for the first time an essential Gothic property. Robert Louis Stevenson's *Strange Case of Doctor Jekyll and Mr Hyde* (1886) ensured this status:

> A great chocolate-coloured pall lowered over heaven . . . [and] as the cab crawled from street to street, Mr Utterson beheld a marvellous number of degrees and hues of twilight . . . The dismal quarter of Soho seen under these changing glimpses, with its muddy ways, and slatternly passengers, and its lamps, which had never been extinguished or had been kindled afresh to combat this mournful reinvasion of darkness, seemed, in the lawyer's eyes, like a district of some city in a nightmare.[13]

Fog literally provides an atmosphere of evil and intrigue here, and thereafter. If the nostalgic version of Victorian London imagines the town covered in a perpetual mantle of crisp, white sentimental snow, then the nightmare version of the century's end paints it permanently swathed in swirling, sinister fog.

Wilde's view aside, there is historical substance to these artistic confections. The Met Office's records indicate that the lowest ever daily sunshine record occurred in the City of Westminster in December 1890, where o.o, I repeat, o.o hours of sunshine were recorded for the whole month. A mournful re-invasion of darkness indeed. Stevenson wrote his tale in Bournemouth in 1885, where he had fled to find a more forgiving climate for the bronchial complaint that eventually killed him. Racked in the agony of constricted lungs, small wonder he imagines a nightmare city cloaked in a suffocating pall that cuts off the light of heaven. As we shall see in the next chapter, it was this darkness (an infernal concoction of coal smoke and Thames fog) that created such an urgent need to let the sunshine into the first decades of the twentieth century.

But fog is not the only element to chill the bones of our culture. As Peter Ackroyd remarks, 'English fiction is . . . drenched in rain.'[14] Rain stops play, starts colds or allows assignations in Austen and Charlotte Brontë; frost and flood assist Hardy's tragic ironies. Both *Wuthering Heights* (1847) and *Bleak House* (1853) are named after houses named after what the elements do to them. *To the Lighthouse* (1927), a founding text of literary Modernism, an avant-garde exploration of desire, loss and narrative form, begins and ends with a very ordinary, quintessentially English concern – whether it will rain the next day. It does, of course, and the aborted trip to the lighthouse creates a profound symbolic lack that the rest of the narrative strives to resolve. 'Ladies and gentlemen, welcome back to England,' declares Stanley Holloway in the final scene of *Passport to Pimlico* (1949). Right on cue, the thunder rolls, the heavens open, and the heatwave ends with the film. Pimlico's restoration to English soil is emphatically signalled with a downpour. That rain kept falling through the next decades, when the Kitchen Sink tradition depicted life sunk in a peculiarly English puddle of misery. If our young men were particularly angry in those days, no doubt the weather had something to do with it.

What a contrast classic American films from the period present to our own signature pieces. Neither *The Wild One* (1954) nor *Rebel without a Cause* (1955) can blame the Californian climate for their youthful rebellions. In *The Graduate* (1967), the first words we hear, as Dustin Hoffman's plane makes its descent into Los Angeles, is that the 'weather is clear; temperature is 72'. Hoffman is another confused, restless young man. But he expends his aimless angst by drifting atop an inflatable in a sun-spangled swimming pool the whole summer through. If you've got to be restless, if you've got to express your youthful frustrations or signal your rebellion, there are worse places than California to do this. (It was never quite the same for me in Croydon.) In *Easy Rider* (1969) two young men take to the road on big choppers from LA to New Orleans: 'In search of America', as the promotional poster for the film explains. But they 'couldn't find it anywhere'. OK, but they found some lovely weather and some pretty far-out drugs instead, enjoying nothing but glorious sunshine for the whole journey. I can't hear the Byrds' music without seeing those choppers slicing through the endless sunlit vistas connecting LA to New Orleans. Sunshine posi-

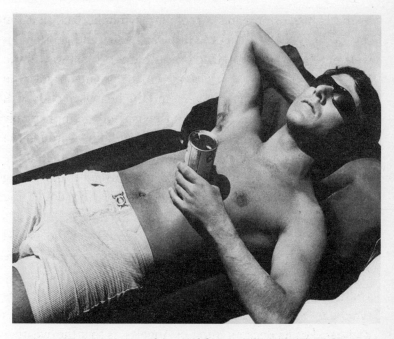

Angry Young Man, southern California style. (Dustin Hoffman
'drifts' in *The Graduate*, 1967)

tively radiates from the screen, making this, not surprisingly,
one of my favourite movies.

So, how were the 60s for us? 'A disgusting time to be alive',
according to Bruce Robinson,[15] whose cult film *Withnail and
I* (1987) offers a very English counterpart to the Californian
version of the last days of that decade. *Withnail* is also set in
1969 and, like *Easy Rider*, shows the disintegration of the
counter-cultural ideal for two young men. But what a different perspective. They are broke, they live in squalor (this is
truly Kitchen Sink drama, turning the yearly washing up into
a gripping tussle with the unknown). It's 'perpetually cold',
and it's only September. So cold they smother themselves with

Vaporub (when they're not considering drinking it) to stay warm. It doesn't stop raining. They also take to the road, holi-daying 'by mistake' in Cumbria, where it's even wetter than

West Coast rebels. Between LA and New Orleans, 1969. (Denis Hopper and Peter Fonda go in search of America in *Easy Rider*, 1969)

London. They wear plastic bags on their feet instead of flowers in their hair. There is nothing remotely swinging, happening or groovy about their 60s (a particularly wet decade in Britain). Piss-headism and pissing down – a very English revolution.

Yet, instead of stifling creativity, grey can actually be a spur. That's Jeremy Paxman's view, when he proposes the 'reason-able supposition that cold wet weather, which forced teenagers to stay indoors in winter instead of going to the beach or skiing, probably has something to do with the country's capacity for inventive rock music'.[16] Paxman, who is on surer ground talking about John Bull than Johnny Rotten, leaves this intriguing idea unexplored. I have formed my own theories.

Whilst the UK may excel at perfecting high-quality pop product, it is also true that the idea of pop didn't originate here. From its origins the story of the UK's relationship with

Wet Coat rebels. Between Camden Town and Euston Station, 1969. (*Withnail and I*, Richard E. Grant and Paul McGann, 1987)

pop has been one of creative modification of sounds, styles and attitudes from elsewhere. From jazz to rock 'n' roll, blues to northern soul and hip hop, pop culture has generally been an exotic importation from a very different clime. There was always something gawky and scrofulous about the early British rockers. The Teddy boys had impetigo to a man, whilst our home-grown Elvis wannabes paled, literally, in comparison. You can quiff your hair, curl your lips and thrust your hips, and attempt an American accent, but something is still missing. What's always missing. Sunshine. Bright days, warm nights, healthy complexions. Down below those borrowed clothes, we were always a shade too white.

There is, however, one supreme example of British rock

inventiveness for which this is an advantage. We might view Goth as an existential recognition of the reality of death. A charity-shop melancholia, parading skeletons at the bubble-gum feast of teenage pop. Or, it might simply be seen as a very pragmatic response to our weather. One of the few truly home-grown pop style cults, Goth has thrived by making a sartorial virtue of a climatic necessity. For the Goths there's no such thing as a shade too white. Their ideal complexion is that of a corpse dragged out of a swamp. I'm informed that Australian Goths have a taxing time of it, and have to employ parasols (black) to maintain the look. Goth draws deep on British romanticism, northern melancholia and the sodden reservoirs of a life lived indoors.

But do we imagine it? Is the weather in the UK quite as bad as we have convinced ourselves and others? On the whole I'd say it was. The Met Office's website endorses this rather depressing suggestion in its overview of our climate. Situated on the north-west extreme of Europe, these isles are dominated by westerly winds bringing 'depressions and their associated weather fronts (bands of rain and cloud)'. However, 'sometimes large, stationary anticyclones . . . act as a "block" to the regular passage of depressions . . . [and] can lead to drought conditions, as rain-bearing fronts are "diverted" around the country.'

So, it's official. Fine weather of any duration is a mistake, a happy accident blocking (like atmospheric Prozac) the 'regular passage of depressions'. Phrases such as 'sunny intervals' or 'sunny spells', beloved by TV weather forecasters, pretty much sum up the outlook of the British weather. As if rain and gloom were the main feature, broken by an occasional interval. A commercial break offering a brief tantalizing glimpse of a better brighter world, before normal service is resumed. A 'spell' of good weather only occurs because some wayward exotic system indolently tarries over these isles, before being chased away by a brisk frigid westerly. But like the dream images of Hollywood and adland, this exotic interval, this 'sunny spell', has cast its magic glamour over us. A teasing taste of it, just enough to distort our expectations and, like all effective adverts, to create unfulfilled desires.

SOUTH FACING

George Orwell is about as English as you can get. His concern with depicting the plight of the dispossessed or the trapped, or imagining far worse to come, led him to paint a fairly dismal picture of his country. It is nearly always winter in his novels, the skies are generally grey, and his characters have holes in their shoes. The Orwellian climate expresses perfectly the outlook of his protagonists. Yet we also catch fleeting glimpses of something beyond these prison bounds – another world that is always bathed in sunlight. In *Nineteen Eighty-Four* (1949), Winston has recurring dreams of something he calls the 'Golden Country', which he discovers to be real when he leaves London for his illicit rendezvous with Julia. He walks through 'dappled light and shade, stepping out into pools of gold wherever the boughs parted'. The air, so different from London, 'seemed to kiss one's skin . . .'

Back in London shortly afterwards, Winston hears a woman singing through the open window of an apartment. He peeps out:

The June sun was still high in the sky, and in the sun-filled court below a monstrous woman, solid as a Norman Pillar,

with brawny red forearms . . . was stumping to and fro
between a washtub and a clothes line.

Such evidence of the indomitable spirit of the 'proles' gives
Winston brief hope that resistance is possible. 'The birds
sang, the proles sang, the Party did not sing.'[17] There is something characteristically British in the way freedom, happiness
and their opposites are depicted here. A sun-filled court
hedged in by a grey, hostile city speaks feelingly of an enduring privation. Yet there is no intrinsic political significance in
what he describes. This is simply the reality of life lived under
these skies.

Sunshine always offers a fleeting glimpse of something
missing from our lives, something that might sort us out,
make us whole, make us happier. We like sunshine because
we believe it is so rare. Our near-unquenchable sense of
solar deficit means disgruntlement, longing or nostalgia, and
unseemly bingeing when the sun does appear or when we
outsource it overseas.

In recent years we've convinced ourselves that we've
become more 'Mediterranean' in our tastes and sensibilities.
We've even essayed a sad pastiche of café society, with
pubs, bars and cafés now optimistically arranging deckchairs
(Titanic-like) on their forecourts, complete with heaters and
windbreakers. One vital ingredient is missing: the weather
and, with it, the nonchalance that attends a predictable warm
climate. That's the reality of 'cool Britannia', and the patio
heater is its symbol. There is nothing cool or sophisticated
about the frantic population of every square inch of urban
scrub at sunny lunchtimes (making parks and city squares
resemble wildlife reserves), or the cascade of beery cama-
raderie blocking pub-side pavements on fine evenings. Mad

dogs and Englishmen will always go out in the midday sun. Making the most of it is built into our hard drives.

Such solar bingeing finds a direct parallel in our approach to alcohol, where yet another (and not unconnected) north-south divide appears to operate. An academic historian of booze, Ruth C. Engs, explains how bingeing has always characterized the northern attitude to alcohol:

> Infrequent, but heavy, drinking to intoxication may have developed among northern Celtic and Germanic tribes in antiquity, because alcoholic beverages were not always available due to variations in the weather . . . If alcohol production was limited . . . a 'feast or famine' situation may have occurred leading to sporadic bouts of heavy drinking to intoxication whenever any alcohol was available.

She also observes how 'people tend to stay indoors during cold damp weather which is more common in the north compared to the sunny south'.[18] Add to this the long, dark winters of the north leading to depressions and relief in drunkenness, and the parallels between solar and alcoholic bingeing start to converge.

In the Mediterranean area, however, 'daily wine consumption with meals' was established in antiquity. Such familiarity is still naturally present from an early age, with children drinking diluted wine at the table. In the UK we grow up with an acute awareness that drink is an adult thing, and usually to 'celebrate' something. Alcohol gets you drunk, so alcohol is for getting drunk with. This carries on into adult life with restrictive opening hours and other sure-fire ways of making us binge. A friend once said, as he knocked back another glass of free wine at some literary do, that he would know he had grown up when he no longer felt compelled to get drunk on free wine simply because it was free. I will know I have grown up when I can apply the same discipline to sunshine (and wine, too, come to think of it).

Sun and booze are two sides of the same coin. 'Feast or famine' and the 'long dark nasty winters' make their presence felt again. And so we binge. There are no rites of spring in this country. Only summer. No gentle, diluted, leisurely ambling into fine weather. As soon as the sun comes out so do we. Down to the beaches, out with the barbies, off with our clothing, and down the hatches with the booze. Cramming a whole summer into a day, because, as we tell ourselves – stretching out in our deckchairs and pouring another drink – it's not here all the time. The sun pulls the cork on life, making any day a 'special occasion' by the mere fact of its presence. Be it a chilled white on a Tuscan terrace, a bevy of beers on the Brava or some scarcely chilled Cava out of plastic cups in a park, booze will always accompany our solar bingeing. They go together, because they are forged in the same mint. The mint of perceived scarcity.

These, then, are the conditions of life lived under these northern skies. The grey and the golden threads that run through our culture and possibly bind us to our ancient ancestors who experienced similar weather and nurtured similar desires.

But is this an accurate picture? When we shed our clothes, stretch out in the sand and bathe ourselves in that golden fire, are we merely reverting to ancestral type as desperate devotees of a fickle and fugitive deity? Is it ancient history, universal biology or acquired habit that compels us to get out there, make the most of it here, or go in search of it abroad? These are the questions that recur throughout the journey that follows. Here are the, often surprising, answers I found.

3

HEALTH

Fear no more the heat o' th' sun . . .
William Shakespeare, *Cymbeline*, IV.2.259

A t first all you see is blue. Then a solar flare splits the screen, and a cicada gives voice to the sun drumming down mercilessly on the supine form lounging by the pool. So opens *Sexy Beast* (2000), a perfect paean to the British love of sunshine. Gal, played by Ray Winstone, is an East End villain who has worked long, hard (and illegally) to prostrate himself before this idol. The film is billed as a love story. And so it is. Gal is a tanned, trunked troubadour hymning the true object of his devotion.

'Oh, yeah. Bloody hell. I'm sweating like a porker here . . . It's like a fucking furnace, that sun . . .' People say, 'Do you miss [England], Gal?' I say, 'No bloody way.' They say, 'What's it like in Spain then?' And I say, 'It's hot, hot, hot. Fucking hot.' 'Too hot?' They say. 'Not for bloody me,' I tell 'em, 'because I love it.'

Of course he does. And he's not the only one.
What could be more natural, obvious and easy than lying in the sun doing nothing? It is done intuitively as soon as the

sun comes out in Britain, and vast amounts of money, time and effort are expended to get abroad when it more characteristically doesn't. Most holidays devote at least a few hours to this activity. It ritualistically declares that the holiday has started, and it is time to really relax.

This is sun-worship as we know it, and as we have known it for some time. A spectacle to make more sophisticated travellers weep, and health agencies shake their collective heads and wonder if they will ever wean the besotted bronzers off this dangerous addiction. Safe Sun campaigns may have been successful in countries like Australia, but we Poms still like to frit. Sunshine, whether domestic or outsourced, is just too precious a commodity to be wasted. And the desirability of a suntan, and the rites of its acquisition, are so deeply ingrained in northern cultures that it may take generations yet to turn the tide of desire.

Perhaps as long as it took to instil it in the first place. For this apparently intuitive activity took quite a while to be acquired and perfected. Before the twentieth century those who could shunned the sun. The fashion was for delicately pale, nearly translucent skin; anything darker was decidedly déclassé. Only the poor were tanned. A bronzed skin betokened lives spent toiling under the elements. Their betters covered their bodies, pulled the curtains, and hid behind parasols and wide-brimmed hats. Like modern-day Goths, they even used cosmetics to make their skin look or remain lighter. *Vogue*'s first British number from 1916 advertised Helena Rubinstein's Valaze Cream. The 'impurities' it claimed to remove included 'sunburn, freckles and discolouration'. 'Burn' and 'tan' meant the same thing then, and 'tanning' referred to the process of preparing leather. Not an attractive image.

By the mid-20s this had changed. By then the poor had

largely forsaken the fields for sunless factories and mines, allowing the genteel to do a complete volte-face. The great outdoors called, and if you were out in the sun you were going to get tanned. But that was OK now. Legend gives the credit for this to Coco Chanel, telling how she accidentally got tanned on a yacht, and started a craze overnight. Coco may have turned tanned skin into a fashion statement, but sun-bathing and tanning had been practised and talked about for a good thirty years before that – just not in a form that was appropriate for the pages of *Vogue*, nor quite how we would recognize it today. The only vestige of this prehistory survives in the phrase 'a healthy tan', which is now considered by killjoys in white coats to be a contradiction in terms.

But a hundred years ago sunbathing and tanning were *only* indulged in for health reasons, to cure or prevent diseases:

Dressed for the beach, Brighton 1900

including those medical science now attributes to the very thing about which it was once so evangelical. Medicine created our addiction to sunbathing, a monster it largely refuses to acknowledge as its own and can't coax back into the lab. The greatest revolution in attitudes to sunshine took place in the sanatorium, not on the catwalk, and it is here that we must go to find the origins of Gal's favourite pastime.

DR SOL

In 1903 Dr Auguste Rollier opened the world's first dedicated sun clinic, at Leysin high in the Swiss Alps. He was convinced that the pure air and bright sunlight could cure a variety of diseases – most particularly forms of external (i.e. non-pulmonary) tuberculosis, which were generally treated, often unsuccessfully, with surgery. His book *Heliotherapy* offers photographic evidence that dramatically proves his point. Pale, pigeon-chested specimens with horrific skin lesions are transformed by a few months of surgical sunning. They sit up straight, they smile, their eyes bright with health and happiness. And their skins – now clear of disfiguration – are burnished the kind of deep mahogany that many a modern sunbather would give their flip-flops to achieve. The bronzed beaming kiddies bathed in sunlight on the balconies of Rollier's sanatoria could easily be so many modern sun-worshippers stacked up by a pool.

But, despite appearances, the two practices should not be confused. What we are witnessing is the birth of a therapeutic method, the establishment of a new science, and the glimmerings of a revolutionary new dawn. It was serious stuff that Dr Rollier was dispensing at Leysin.

Fun and sunshine. Rollier's sun cure at work

On arrival, Rollier's patients didn't just strip off and dive in like so many sun-starved Brits on the first day abroad. 'During the first few days rest in bed indoors is imperative, and only gradually is the patient allowed to get accustomed to the open air through judicious use of fan-lights, windows and doors.' This gradual process of acclimatization takes about two weeks, and only then were his patients allowed to dip their toes into the 'sun baths'.

You start with the feet, for exactly five minutes, repeated three times daily for about three days. (Isn't this fun?) Next the legs, for similar treatment, followed by the trunk, and finally the head. Full solar immersion takes about five weeks, and only then would they start to treat the critical area. Precise, prescribed, protracted, Rollier's 'insolation' was conducted on the strictest scientific and therapeutic principles. As he asserts: 'No other method requires such a meticulous and strict individualization and exact adaptation to different cases as the sun cure does. A strict adherence to these rules is indispensable, since the success or failure of the cure depends on it.'[1]

From being feared and avoided, sunshine became highly valued. Sunlight was of course as free as air, but this new sun, the healing sun, at first belonged to medical science. Rollier insisted that Alpine sunshine and air were of a specific quality and an essential part of the cure. But sunshine could only be localized up to a point (that point being about 1,300 metres above sea level), and so the precision of the process was

Beach, towel, relaxation? Not a bit of it.
The sun had important work to do first

essential. Rollier was not just turning sunshine into moon-shine here, sophisticating a free resource to hoodwink the desperate. He was no doubt sincere in his wish to alleviate suffering, and dedicated some of his thirty-five sanatoria specifically to the rachitic or tubercular poor. The strict con-ditions he set were more an indication of just how alien the concept of sun exposure was at the time, and of how much his more genteel patients needed coaxing out from the shade, and to accept the consequences – the tanned skin formerly the brand of the peasant.

Layers of prejudice had to be peeled away along with their clothes. And so the process was rarefied into a therapeutic mystery of extraordinary complexity. Rollier's most fervent English disciple, Dr Caleb Williams Saleeby (who crops up often in this story), explains that the ultraviolet rays that effect the cure also 'cause the amusingly extreme pigmentation of Rollier's patients'.[2] Such sights were evidently still cause for comment. The overthrow of prejudices could perhaps only have happened with people who were sick, and so less particular about their appearance. Tubercular or tanned? The choice is yours . . .

A 'healthy' tan meant exactly that: proof that the rays had worked their curative magic. Health appeared to radiate from these rejuvenated bodies, and after a while these rays extended their influence beyond the sanatorium. A new ideal rose with the dawn of the twentieth century. A new age of bronze had arrived.

Rollier was not alone in espousing the therapeutic or hygienic properties of sunlight. As the new advocates often pointed out, the benefits of sunlight and air had been a central part of Hippocrates' theories (fifth–fourth century BC), but had to be rediscovered in the nineteenth century, when nature worship and industrial pollution encouraged a re-examination of these resources. In 1903 the Danish physician Niels Ryburg Finsen was awarded the Nobel Prize for his use of artificial sunlight to cure *Lupus vulgaris* (tuberculosis of the skin). And John Harvey Kellogg (of breakfast cereal and *Road to Wellville* fame) claims to have employed sun baths 'under medical supervision' at his famous Battle Creek Sanatarium as early as 1876.[3]

In 1893 Kellogg exhibited his first 'incandescent light bath' at the Chicago Exposition of that year. A number of these contraptions (the forerunner of the modern tanning capsule, but looking like a cross between a church organ and a crematorium oven) are proudly displayed in his *Light Therapeutics* of 1910. According to Kellogg, sun baths work

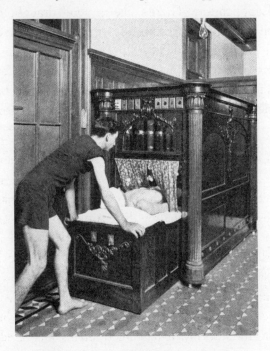

The forerunner of the tanning booth.
John Harvey Kellogg's electric light bath,
first exhibited in 1893

'wonders' for gout, rheumatism and certain forms of tuberculosis, and can effect cures on patients with eczema in 'cases which seemed quite hopeless'. Both Kaiser Wilhelm II and

King Edward VII owned Kellogg light cabinets, the latter reputedly installing them at Buckingham Palace and Windsor Castle.

The Road to Wellville by T. C. Boyle (1981) depicted Kellogg as a bit of a humbug: repressed, right-wing and clearly bonkers. And yet, he appears to have been a largely unacknowledged pioneer in the heliotherapy business, displaying a quite advanced understanding of the science of light and its clinical application. There's also a good draught of Yankee common sense in his prescriptions, especially when compared with the prissy precision of Rollier's rarefied regimen. His sun cure was part of the holistic regimen of diet, exercise and enemas for which his Battle Creek Sanatarium has become an object of ridicule. Kellogg explains his method here:

> The sun-bath is best administered in an outdoor gymnasium, provided with suitable couches, a sand bank, and other appliances. Several patients of the same sex may be treated at once in such an inclosure, the demands of modesty being satisfied by the scantiest of bathing attire. Male patients commonly wear very small trunks, jockbands, or narrow loin-cloths . . . When it is desired to expose the entire skin surface, – and this is always an advantage, – tight screens may be placed about the patient in such a way as to protect him from observation while permitting the sun's rays to fall directly on his uncovered body.[4]

With their skimpy attire, comfortable couches and even sand banks, we can easily imagine these patients as our counterparts, who might actually have enjoyed their treatment. If they did, it isn't mentioned. Pleasure was the last thing to be associated with the 'sun bath' in its experimental days, and certainly not as an end in itself. The 'sun bath' emerged initially by analogy from the therapeutic bathing and water-taking culture established in the eighteenth century. The Battle Creek Sanatarium started out as a hydrotherapy institute in 1866, and

Sun baths at Kellogg's famous Battle Creek Sanatarium. The great man himself is just in the picture on the left

the 'sun bath' was but the latest product line in the burgeoning nature cure industry. You took it under strict instruction, and to cure a specific ailment. It would be some time before the pleasure of sun-*bathing* (the experience itself) was emphasized. The sun had a lot of hard work to do before then.

THE CHARGE OF THE LIGHT CRUSADE

It was not long before sunlight was being advocated not just to cure, but to prevent illness, and even to improve the health of the general population. This need was deemed to be particularly acute in England, the greatest and most polluted industrial centre in the world. As Dr Caleb Williams Saleeby asserted in the preface to the first English translation of Rollier's *Heliotherapy* (1927): 'the real meaning of Dr Rollier's work is not clinical but hygienic . . . the restoration of sunlight to our malurbanized millions, now blackened, bleached and blighted in slums and smoke.' Saleeby made this ideal a positive crusade. He founded the Sunlight League and its journal *Sunlight* in 1924, and devoted his whole life's work to this cause.

For Saleeby, sunshine was more than an atmospheric condition or a curative agent. It was a veritable symbol of salvation. His great cause was to 'pursue the truth that light, or certain kinds of light, may be directly destructive to our enemies of various kinds'. By these he meant diseases such as rickets or tuberculosis, the latter being dubbed by Saleeby 'the prince of the powers of darkness'.[5] Such language befits the pulpit or pulp novel more than the clinic or hospital ward, evoking the lurid hues of the urban Gothic fictions of the *fin de siècle*. Eighteen eighty-five, the year Stevenson wrote *Dr Jekyll and Mr Hyde*, which has fixed this image of darkest London in our imaginations, was also the year the Lever Brothers first trademarked Sunlight Soap.

Between the wars, sunshine didn't just acquire a layer of history, it was history. Or rather it absorbed and reflected the desires and attitudes of the era. As the critic Paul Fussell has noted, after the Great War 'something new and recognizably "post-war" surfaces in British intellectual and imaginative life'. There was a restlessness and discontent with British life which sent countless writers and intellectuals into self-imposed and sometimes peripatetic exile. Their collective lament was 'I hate it here', and looming large on their dismal landscape of annoyances was the weather.

It sometimes seems that it is only after the war that the British weather becomes a cause of outrage and a sufficient reason for departure . . . Before the war one had been rather proud of the fogs and damps and pleased to exhibit staunchness and good humour in adapting to them . . . But after 1918 it is as if the weather worsens to make England all but uninhabitable to the imaginative and sensitive.[6]

The weather didn't worsen – the interwar years boasted a number of exceptional summers and only one severe winter (1928–9) – it's just that the sun was newly valued, newly desired. You get a sense of the sun being discovered for the first time.

This new god rose with the early decades of the twentieth century and reached its symbolic zenith in the interwar years. It had its devotees, its temples – the new lidos and bright white flat-roofed Modernist homes with balconies modelled on sanatoria; its dedicated vestments – the new streamlined bathing costumes; and the sun as symbol was everywhere. Health, hope, freedom, youth and modernity radiated from it. It became the design motif of the era, appearing on everything from cufflinks to stained-glass suburban windows, from garden gates to wirelesses.

But smog-suffocated Britain was not the only place to feel the urge to pull open the blinds on the past and let the bright future flood in. Consider Saleeby's comments about the 'prince of the powers of darkness' and the importance of sunlight alongside the natural history of the vampire. Or at least the popular incarnation of that figure.

As everyone knows, vampires hate sunlight. The biggest joke of the spoof Dracula film of 1979, *Love at First Bite*, was the casting of George Hamilton, famous for his dedication to sunbathing, as the supposedly photophobic Count. There are a few hints of this aspect of vampirism in Stoker's *Dracula* (1897). The team that unites to destroy the Count dubs itself the Crew of Light, while Professor Van Helsing points out that Dracula's powers of metamorphosis are restricted during the day. But Dracula does appear during daylight in Stoker's novel. Whilst photophobia is now an established 'fact' of vampire mythology, it is not found in any folkloric sources. These vampires generally had ruddy complexions, and didn't appear remotely bothered by sunlight. Indeed, Robert Louis Stevenson's short story 'Olalla' (1885) even featured a vampire who spent most of the day lazily 'following the sunshine' around the courtyard of her ancient Spanish *residencia*.

Sunlight phobia is a modern invention, originating with the first vampire film, Friedrich Wilhelm Murnau's *Nosferatu* from 1921. The medium of light, the art form of the future, used sunlight to vanquish the atavistic threat to health and modernity. The German director's film is a thinly veiled adaptation of Stoker's novel, and a flagrant breach of copyright. Stoker's widow pursued Murnau through the courts, and the prints should have been destroyed. But a few escaped, and the film is now considered an Expressionist horror classic.

Nosferatu relocates the action in Germany, and explicitly associates the arrival of the vampire – a nasty ratty bald critter played by Max Schreck – with disease. A plague follows his arrival at Wisborg, and abates with his death: his death by sunlight. The final scene depicts the heroine nobly sacrificing herself to the vampire's lust to delay him until sunrise. The daylight that floods through the window dissolves the vampire before our eyes: a *deus ex machina* of hygienic redemption for this 'prince of the disease of darkness'. The trick caught on and is now a standard of vampire flicks and folklore. Murnau's other masterpiece was called *Sunrise* (1927).

But Murnau's innovation was eloquently expressive of a specific time and place – Germany in the 1920s, where secular sun-worship was more advanced than anywhere in the world at the time. The sun was central to a number of cults and movements from the period. And nowhere so starkly as in the *Freilicht Park* and *Nacktkultur* (Free Light and Naked Culture) movements. Started in 1903 in Hamburg, places dedicated to nude sunbathing for health became increasingly popular between the wars. In the history of sun-bathing the Germans got there first, stripping off their kit and lapping up the sun long before the Brits had emerged from their bathing machines and mackintoshes.

Stephen Spender recalls how the sun was 'a primary social force in this Germany', and how 'thousands of people went to the open-air swimming baths or lay down on the shores of the rivers or lakes, almost nude, and sometimes quite nude . . . During their leisure, all their powers of thought were sucked up, absorbed into the sun'.[7] Interwar Germany was a Mecca of physical culture and hygiene, and the Shangri-la of sunbathers and experimental nudists from around the globe. By the time of the first International Nudist Conference in Frankfurt in 1930, it was estimated that there were three million practising nudists in Germany.[8] When the English translation of Hans Suren's *Man and Sunlight* appeared in 1927 it was referred to by a reviewer as 'probably the most outspoken exposition of Naked Culture . . . which has yet been printed in the English language'[9] – although it might look to us now more like a scary proto-Nazi blueprint for racial supremacy, displaying the flower of German youth, bronzed, buff and butt

Efficiency through health.
German sun-worshippers look
to the future (1930s)

naked, stomping through the flower-strewn meadows of the Fatherland.

The nudists were the first modern sun-worshippers, the first to embrace sunshine out of choice rather than medical necessity. For a long time 'sun-bather' was pretty much a euphemism for nudist. One of the first British societies for social nudity was called the Sun Bathing Society, and there was also a British Sunbathing Association, and a National Sun and Air Society in these formative years. Most of the private grounds dedicated to clothes-free living were called 'sun clubs', clearly indicating the idol around which their activities revolved. Nineteen thirty-one was an important year for the movement, with a number of these clubs opening in the south-east, for stripped-down sunbathing and physical exercise on the German model. That was the plan. But, as so often happens in the UK, the weather had other ideas. That year turned out to be a particularly abysmal summer, and, as a visitor to one of the 'sun' clubs recorded: 'Members of this colony sat in vain through all of June and most of July waiting for the sun to shine. They were rewarded with heavy mists, damp fogs and torrents of rain, which made only the briefest exercise possible . . .'[10]

But we Brits are a plucky lot, and hope must spring eternal in the nudist breast. And so, their optimism undampened, they stuck at it, and soon other sun clubs were exposing their members to the elements. The movement is thriving, and many of the clubs and societies dedicated to social nudity today are still called sun clubs, and use solar symbols in their logos. The story of sunbathing as a social activity relates an heroic struggle fought on our behalf by dress reformers, great-outdoorists, and those who simply wanted to go naked in public. They fought for our freedoms; they made Gal, lying in somnambulant oblivion in his thong, possible. Their story deserves to be told.

BUT, NAKED; OR WHAT THE NUDISTS DID FOR US

The founding ethos of nudism was health. From taking sun baths to cure specific ailments, to embracing it as a hygienic principle or way of life was but a short logical step. Sunlight, along with fresh air and exercise, were the Holy Trinity of a veritable religion of physical culture which emerged around the turn of the twentieth century, whose cathedral was the Great Outdoors. The editorial of the May 1920 edition of the (as yet pre-nudist) *Health & Efficiency* finds its editor full of the joys of spring, and waxing evangelical from his deckchair pulpit. As he explains, this month he is in the English countryside, which encourages him

> to speak to you of nothing but fresh air and sunshine. To me they mean so much more than oxygen and heat rays. Fresh air to me means the abolition of all that is stagnant in life and the establishment of cleanliness and sanity. Sunshine means joy and happiness, and an annihilation of ignorance and disease. Fresh air and sunshine mean phys-

ical, mental, and moral Health, Cleanliness, and Joy. So you, my comrades, when you step out into the fresh air and sunshine – as I hope you will on every possible occasion – try to think of the larger aspect which I have hinted at here, and you will know that there is a real gospel in Physical Culture.

Amen to that. As he warms to his theme of 'Sun Bathing and Air Bathing for Health', we catch the first editorial flashes of an idea that would later dominate the pages of his magazine:

> But now that the warm weather is upon us we can take our-
> selves to some secluded spot and enjoy the perfect freedom
> of the natural condition. The beating of the sun's rays on
> the naked body makes for Health, and the soothing fresh
> air will purify the skin and ventilate the blood . . . sun
> bathing and air bathing are fully recognized for their cura-
> tive value and their stimulating effects.

As we can see, the metaphor of sun-worship is not inappropriate, giving the high-flown religiosity associated with this
'gospel' of the healthy body. This also encouraged the belief
that nudists were a bunch of sandal-wearing, vegetarian, tee-
total cranks. Certainly a far cry from the image of Coco
Chanel on her yacht.

True, there were a few cranks early on. And none crankier
perhaps than Edward Carpenter. Carpenter was a pioneering
apologist for homosexuality and sandals (he appears to have
been responsible for introducing this badge of alternative
lifestyle into Britain). His *Civilisation: Its Cause and its Cure*
from 1889 called for a return to a closer harmony with Nature.
As he reasoned: 'We cover our bodies up, and have become
all mind . . . The worshipper must adore the Sun, he must

saturate himself with sunlight, and take the physical Sun into himself.' Man must 'cure' himself of civilization. At such a time, Man will stand 'uncovered to the Sun, will adore the emblem of the everlasting splendour which shines within'.[11] A similar mystical bent characterizes the first British nudist club. This called itself the 'Moonella Group', and was founded in 1924 as an active cell of the English Gymnosophist (naked wisdom) Society. They met in private grounds in Wickford in Essex for two summers wearing nothing but sandals and brightly coloured scarves and addressing each other by such monikers as Chong, Flang and Zex. Their standard greeting, stipulated in their constitution, was 'Blue Sky'.

But the new sun-worship had its devotees towards the right as well as on the bohemian fringe. Saleeby, the sunlight crusader, was also a fervent eugenicist. When he wrote the preface to the English translation of Hans Suren's book *Man and Sunlight*, the two concerns came together. Saleeby hoped that its appearance in England would 'serve to enhance the physique and the "physiological righteousness" of the young manhood and womanhood upon whom . . . depends the destiny of our country and of the ideal for which she stands'.[12] The first published propaganda for nudism in Germany, Heinrich Pudor's *Naked Mankind: A Triumph Shout of the Future*, from 1893 was openly anti-Semitic. Pudor called for a natural 'airing' of the Aryan people, in the removal of its clothes. He looked forward to the day when European skin would be 'Brown as a bread crust, velvety red as a peach'.[13] And, of course, the Fatherland cleansed of racial impurities. But it's not so neatly polarized. Nudism was about 'health', and unfortunately so were a lot of ideologies floating around at the time.

Given these rather inauspicious beginnings, I wanted to discover when and how 'sunbathing' as we might now recognize it – a semi-clothed, non-mystical, apolitical, social (rather than medical or hygienic) pastime that people *enjoy* – came into being. This meant going back to the origins of the nudist movement, and poring over the complete run of the longest-established British naked culture magazine, *Health & Efficiency*. Which in turn meant entering the infamous 'Cupboard' of the British Library.

As the principal copyright library in the UK the British Library is obliged to hold a copy of most books, magazines and newspapers published in Britain. That includes magazines for gentlemanly relaxation. Porn, in other words. The Library has to handle such material carefully, and it should never become the haunt of undesirables, glutting themselves on this cornucopia of free smut. It's probably fair to say the Library was never at the vanguard of the permissive society, and took a pragmatically conservative view in its classifications. The chief classification clerk could not be expected to take more than a cursory glance at the material, so as not to distract him in his job or undermine his morals. Anything that looked dodgy was kept under lock and key, given a 'Cup' prefix, and treated with kid (or maybe, boxing) gloves.

This seemed only right and proper when the Library constituted the holy of holies of that neoclassical temple of culture, the British Museum in Bloomsbury. In those days you had to read Cupboard material on a special table, directly under the vigilant gaze of a senior male librarian. This table appeared to be special by design as well as location. Whilst most of the tables around it had panels underneath (I can only assume to stop men ogling the limbs of female readers), the tables reserved for items from the Cupboard were open.

With such material, it was more important that everything, especially the hands, were in full view.

I assumed this quaint institution had been left behind when the Library took up residence in the efficient, bright, modern and welcoming building at St Pancras in 1997. But even the early issues of *Health & Efficiency*, when it was called *Vim*, still had a Cup prefix, and were still subject to the same institutionalized stigma of old. To consult 'special material' one must go to a special counter, sit at a special table, and get special looks from the librarians when you nonchalantly ask for your books. (Just what expression does a serious scholar wear?)

And so there I sat for what seemed like weeks, at my special table: flicking through pictures of Edwardian strong men in leopard skin loincloths with preposterous moustaches, posing as Samson or Achilles; or through homilies on the perils of self-abuse; the proceedings from the Association of Honour or the Purity League; alarmist debates on the Race Question; regular features by Robert Baden-Powell on the principles of the Scout Movement; articles on fruit and constipation, fresh air and haemorrhoids, auto-intoxication by meat, alcohol or tobacco. In short, the most earnestly moral, most muscular Christian propaganda imaginable. And not a whiff of sauce for decades. I was tempted to order up some vintage *Razzle* to spice up my moral fibre diet, and at least justify the moral opprobrium in which I was bathed at Smut Corner of the British Library.

I had reached the early 20s, and still no signs of nudity. It was only the regular column on physical jerks by a gymnast called Ed Shufflebottom (too good to be true) that enlivened my ordeal and kept me going. But what was gathering apace was a near-obsessive concern with sunshine. Pretty soon the sun was everywhere, serving as the singular most popular symbol of health. Products from prunes to chest expanders, bicycles to health resorts adopted the sun as their patron saint, and employed it in their iconography and advertising. The sun illuminated a new path to health, a road thronged with ramblers, hikers, Boy Scouts and campers, all enjoying themselves in the great outdoors, but all fully clothed. I was beginning to despair of anyone getting naked, when suddenly there were rustlings in the undergrowth of leafy England, as the nudists started to break cover and join this happy healthy throng.

Tentatively at first came letters from 'Gymnosophists', attempting to form societies. It appears there were numerous enthusiasts out there who were privately indulging in naked solar exposure and were keen to emulate the German model of doing it socially. All undressed but with nowhere to go, it was time for British nudism to get organized. By the end of the 20s, 'sun clubs' were being formed, the models in *Health & Efficiency* had shed most of their clothes, and the stigma attached to my vigil at the special table was at last partly justified.

But only partly. Early nudism was not only not sexy, it was insistently unerotic in its look and tone. The mainstays of public school orthodoxy – muscular Christianity and the Hellenic ideal – were invoked to support the early nudists' bid for respectability. The early nude studies which became a regular feature of the reborn *Health & Efficiency* from about 1930 (encouraging a new readership interested in neither health nor efficiency) clearly attempted to live up to this ideal. Models resembled marble statuary, sanitizing any suggestion of sexuality through the formality of their classical poses and the airbrushed absence of genitalia. The titles

One of Britain's longest running sun clubs, 1935

of these artistic studies underlined their high seriousness and artistic pretensions – Aurora, All Hail the Dawn, Behold Apollo in His Glory, I've Got a Lovely Bunch of Coconuts.

Nudist propaganda looked to ancient Greece as its inspiration and ideal. The reality, as encountered in the pages of the early *Health & Efficiency*, was very different. The stagy show-case nudes might have resembled living statues, but the amateur enthusiasts as preserved in film or anecdote are more recognizably English. It is difficult not to feel a deep sense of shame when comparing the English with the German images from the time. Not out of prudery, but at the simple unaesthetic amateurishness of it all: the Germans, buff, bronzed and statuesque in choreographed classical formation in scenes of Alpine sublimity, as against a motley crowd in big pants stumbling through physical jerks in a field near Harrogate. None of the Sun Bathing Society appears to be particularly tanned or enjoying themselves. Scrawny specimens doing awkward things in dreary locations – more Charles Hawtrey than Charles Atlas.

The nudist sun clubs were understandably cautious about the legality of their operations, and took pains to deflect moral outrage at their supposed liberty. They stipulated that males should always be accompanied by their legitimate female partners, and not wolfishly preying on the disrobed damsels within the high-fenced enclosures. Applicants were screened for their moral bearing, whilst at a sun park near Croydon, 'members are under strict supervision of a committee of family men and women, which includes doctors and clergymen'.[14] They must have been queuing round the block for that particular pastoral duty.

The British Sun Bathing Society put on a public demonstration of how it's done in 1932

But such measures were not mere fig leaves. There appears to have been a genuine sincerity in this ethos. When introducing Suren's book in 1927, Saleeby quoted the Dean of St Paul's, a known supporter of the sunlight crusade, to whom he had sent a copy in proof:

> The book will 'do good', as the Dean of St Paul's predicts, because it teaches a much-needed physical discipline, a genuine asceticism, not for the denial but for the enrichment and ennoblement of life.

The appearance of homilies against the perils of self-abuse in the pages of a magazine that was no doubt employed to assist that very purpose, is thus not as bizarre as it might at first appear. Early nudism was not just not immoral, it was painfully and, to us now, comically moral in its ethos. Like Boy Scouting but with nowhere to sew the badges.

RUDE HEALTH

It is easy to snigger at the stodgy earnestness of these early sun-worshippers; but by fighting for their own freedoms they made possible a lot that we take for granted in our relationship with the sun. Their tireless propaganda for clothes-free sun exposure helped make bathing machines and woollen one-pieces a thing of the past. In fact, before the 1930s, there was no real distinction between 'sunbathing' and 'nudism' either within or outside the movement. Given what constituted acceptable and even 'legal' beachwear at the time, complete nudity and what is nowadays standard beachwear might just as well have been one and the same thing. For many it was. A newspaper from May 1925 reports how Bournemouth was strenuously objecting to sunbathing on its beach. It pointed out how beach attendants had powers to prevent anyone sitting on the beach in their bathing costumes. Bathers must 'walk straight into the sea and straight back to their bathing tents'. The Twenties were not yet Roaring down in Bournemouth, evidently.

The term nudist didn't really appear before about 1933; before that 'naked sunbathing' was used to denote the militant exponents of a practice that was by then gathering broader acceptance and popularity. Nudists were sunbathers with attitude. And yet the first rule of the Sun Bathing Society was that 'proper Sun-Bathing costume should be worn' at the sun clubs under its auspices.[15] The Society adopted a softly-softly approach to establishing freedoms for nudists, and sought to be the acceptable face of the movement, organizing open days with displays of physical culture in the semi-nude. And while it later relaxed these rules, and allowed discreet total nudity in its clubs, at first it was simply fighting for what is now *de rigueur* for sunbathing – skimpy slips to allow fuller solar exposure. Modern sunbathers are merely clothed nudists, and owe a lot to these radical heliophiles.

And yet, the very militancy of the early sun-worshippers also removes them from what we recognize today. Suren's opening address to his ideal reader reveals how a dedication to sunbathing could verge on the fanatical for its early acolytes: 'Greetings to you, you who are sun-lovers! You bear ardent longings in your hearts! Longings after warm sunshine, blue skies, light and nature.' So far, so familiar. But he continues: 'Exultingly do you rejoice when the smallest beam of light gilds the altar of your longings. Out of passion for sunshine springs the noble shrine of loftiest idealism . . . Hail to you, you who love sunshine and light, avidly, openly, fervently!'[16] Lazily, indulgently, indolently? Not a bit of it. And so on for another two hundred pages.

Physical culture was a characteristic feature of the nudist movement both in Germany and the UK, making early sun-worship look like hard work. Flick through nudist publications from the first thirty or even forty years of the movement and you'll lift the lid on a buzzing hive of communal activity – enthusiasts forming naked human pyramids, swinging clubs, leaping about or playing tennis. Always doing something. This is partly to reinforce the morality once again. We may see naked men and women together, but not lying there

Sunbathing was hard work in
the pioneer days

indolently in available poses, there's no time for that. Hup hup hup. And when we've finished with the medicine ball, Clarence needs a hand building the new latrines. Come on, you slackers. Last one there's a mere sunbather.

You'd think a fair-weather sun-worshipper is wholly tautological. But not according to this apologist, whom the washout of 1931 obviously put on the defensive: 'Naked and unashamed, laughingly let us run, dance, swim and sing in all sorts of weather. Vagaries of climate, cold winds and even snow need be no obstacle to a fit, hardened, unabashed and sensible people.'[17] OK. You first. But this hardiness is also a sincere expression of the principles upon which sun-worship was founded. And it became a characteristic feature of social nudity once sunbathing became popular with the general population who had no intention of removing all their clothes, and certainly not when it's snowing. With the emergence of dedicated nudist 'colonies' in the early 30s, sunbathing and nudism (or naturism as it would soon be called) started to have distinct identities.

Naturists were the pioneers, purists and high priests of sun-worship. Whilst others might flock to the coast and lie there toasting, the naturist stood apart, and worked hard to maintain distinctions. And so, whatever you do:

> Don't lie on the beach! No matter what the sun's like, don't lie in it! . . . But what about all those rows and rows of sun-bathers on the sands? They're wrong. Don't lie down. Stand up, move about, keep active. This doesn't mean playing violent games *all* the time, and exhausting yourself; just keep moving, be sufficiently active to keep your circulation going. Let the air get to your skin as well as the sun.[18]

Real sunbathing requires effort, and should be carefully regu-lated along strict scientific principles. As he advises: 'Don't forget to wear your watch.' What? And spoil my perfect all-over tan? 'You'll need it for timing the exposure – the expo-sure, remember (and don't you ever forget), that has got to be strictly progressive, a bit more and a bit more each day.' Whatever you say. Now, would you mind moving over there to do your routine, you're blocking my sun.

But what of those rows and rows of sunbathers, lying lazily on the sand? Through the stridency and sniffiness of this naked purist's prescriptions we catch a glimpse of a much more familiar scene. The naturists might still be leaping about and maintaining the hygienic purity of their precise solar exposures, but it looks like those outside the sanatoria and nudist colonies were freely lapping it up – were doing what we've always done whenever the sun shines, and that's to make the most of it. Maybe not quite always. By 1940 we evidently were, but it would appear that what is now a sacred principle of a sun-starved population had to be learned. Habits we now take for granted had to be instilled in us. We needed persuasion, example and opportunity to turn us into the nation of sun-obsessives we are today.

MAKING THE MOST OF IT

Sunbathing started to emerge as a health fad and leisure pursuit in Britain in the mid-1920s. A run of hot summers must have played its part in the new solar awareness. If nothing else, the summers of 1922, 28, 29 and 30 ensured that people acquired new habits, experimented with new freedoms, and ultimately got a taste for it. The fashionable were of course enthusiastic early adopters. If medicine claimed it did you good, the smart set decided you looked good too. As Britain sweltered in the summer of 1928, a *Daily Mail* society piece of 16 July remarked how a crowd including royalty, actresses and aristocrats had turned Bray-on-Thames into 'an amazing reproduction of Deauville'. As it enthused:

> Pretty girls in striped red and white skin tight costumes reminiscent of Venice sat opposite sun-burned young men with towels on their shoulders, while champagne corks popped and the soft croon of dance music floated out from the restaurant . . .

The sun brings a taste of the exotic to the Home Counties, allowing the rich to show off their newly acquired habits and their newly fashionable complexions. The following year brought another heatwave and another new fad: 'Bathing costume tea parties'. As the *Daily Express* reported on 18 July: 'Women with gardens have discovered a new way of fighting the heat wave . . . Parties sit and lie about in the garden, scorching themselves and revelling in sun without restriction. The fashion has spread like wildfire.' As a Hampstead lady declared, 'I have never enjoyed a summer so much in my life. Instead of tiring myself and paying calls or behaving formally at stuffy tea parties, I just tell my friends to come along here and make themselves at home in the garden.'

Punch, of course, poked fun at these new enthusiasms. A cartoon from 1928 asks: 'Why go abroad when, exercising

Punch pokes fun at the new fad for sunbathing, May 1928

a little imagination, it is possible to revel in the delights of Venice – or the sweet dalliance of the Lido?' A group of aesthetes and bright young things look decidedly out of place on a crowded British beach. They are lying down, scantily clad, and enjoying cocktails and cigarettes with their sun bath. Around them sit the ordinary folk, clinging stolidly to the beachside protocols of the past. They are older, fully dressed and either standing up or in deckchairs. They stare at the new exotics with incomprehension. Two worlds collide on Margate sands.

Was *Punch* correct, and sunbathing a fad adopted only by the leisured or eccentric? To some extent. The flesh may have been willing, the sun may have been shining, but law, custom and habit presented a few obstacles to those who wanted to try it out. As we saw earlier, some seaside resorts restricted what could be done or worn on their sands. Brighton allowed, even encouraged, sunbathing on its 'Beach of Brown Men' from at least 1925. On 23 July that year the *Daily Mail* reported: 'In this morning's scorching sunshine dozens of men, young and old, lay basking on . . . the free bathing beach – wearing in many cases no more than a towel about their waist. Some had skin so brown they looked like Red Indians.' It helped that three times mayor Alderman H. Carden was an enthusiastic sun-worshipper, 'who gained for the town, against much opposition, the right of sun-bathing on this beach'. Brighton may have lived up to its reputation as a trendsetter, but Bournemouth clung fast to its claims to gentility and decorum, and banned sunbathing until the early 30s.

The beach has ever afforded unique freedoms. Neither 'land nor sea, neither nature nor culture', but partaking of both, it is the place where boundaries are crossed.[19] It was where we first learnt to submit our bodies to the waves, and later to disrobe and expose them to the sun. When we are still drawn to the coast in hot weather it is not just that it offers cooling breezes and a more pleasant environment. Does it? Half the day spent in a traffic jam, the other on a noisy crowded beach when we have a nice garden, patio, balcony or park to sunbathe in? It is perhaps also following a strong ancestral call to the sacred space where we first surrendered ourselves to the sun. When a near riot ensued at the Welsh Harp Reservoir in north-west London in the warm summer of 1930, with locals attempting forcibly to evict a group of naked and near-naked sunbathers, it was not just about what they were baring, so much as where they were baring it. They had strayed out of the space sanctioned for public exposure, and a mob enforced the accepted boundaries.

This popular fixation with the seaside as the exclusive site for sunbathing caused concern among its more evangelical exponents. In the *Evening Standard* of 27 July 1928 Philip Page claimed that because Britain is an island 'we tend to associate bathing with salt water and don't really go elsewhere', and then only on the annual holiday. 'Thus for fourteen days only are English bodies exposed to the rays of the sun. Probably for less, for with the English climate it is likely that some or even all of those days are sunless.' Page pointed out that the 'sun bath' after the dip was as important as the swim itself, and contrasted the cautious Brits with the fervent Germans. There the 'sun bath is not just an incidental episode in a short summer vacation'. They know how to make the most of the sun, even using their lunch hours to catch some rays. In Britain, where 'we do not see enough of the sun . . . we do not make enough of the sun when we do see him'. This was cause for concern.

From the perspective of some eighty years this is a revelation. If the Germans started the trend, the British have learnt to catch up, and are now neck and neck, towel to towel the Med over. We now display a near-demented devotion to sunshine, and an heroic determination to make the most of it every time it appears. 'Phew, what a scorcher!' declares the tabloid, and devotes at least a spread to recording a nation determined not to waste a drop. But, outside nudist and fashionable circles, it looks as though this is an attitude we had to learn, and were actively encouraged to adopt. Health experts displayed an extraordinary faith in the virtues of sunshine, and made it a popular propagandist cause. They looked enviously and then nervously at German pre-eminence in sport, health and sun-worship, and called on us to emulate them. The race to secure a place in the sun had a different meaning then, and was no laughing matter.

TOP UP

Sunbathing obeyed 'economic' principles, in a number of senses. Whilst the leisured and fashionable were both more willing and able to make the most of the sunshine in their private gardens or clubhouses in a heatwave, or could afford to flit off abroad when the British climate followed its more characteristic pattern, the landlocked, factory-farmed masses had an uncertain two-week annual window to get their 'quota'.

This idea of a quota of sunshine indicates the other, closely related sense of the sunshine economy. Those who promoted the healthiness of sunshine tended to represent it as a substance you had to fill up on if you wanted to remain healthy and happy. You built up your resources in the summer, which you drew on to get you through the winter. Those who popularized sunshine as a medicine were generally unspecific about what its 'tonic' properties consisted of. It sufficed to know that the stuff was good for you, that there was a finite and unpredictable supply, and it was your duty to make the most of it when it appeared.

Top Up and Top Off became the rallying cry of the new Light Brigade, as medical experts lent their support to the need for dress reform in promoting the health of the nation through greater solar exposure. This need was particularly acute for men. Whilst women were more ready to adopt the new summer fashions that steadily increased exposure throughout this sunny decade, male sartorial codes moved at a snail's pace by comparison. We're not talking about hot pants on High Holborn here, but the simple right for men to wear soft collars or go without waistcoats during hot weather.

Men in 1925 essentially dressed as their fathers or even grandfathers had done before them. But that was no longer appropriate for these more relaxed and enlightened times. And so, in June 1930, the *Daily Express* (hardly a radical organ) initiated a Making the Most of It campaign. It was another hot summer, and the paper claimed men were absurdly overdressed. So it declared an all-out 'Sunshine Dress War', and invited medical experts to enter the fray. Sir Bruce Bruce-Porter, the eminent physician, pledged his support to 'the movement against wasting sunshine', but despaired of convincing the present generation of men. As he explained:

> our men are hidebound by convention . . . the great mass would rather lose the benefits of sunshine than brave a revolution in fashion . . . We must propagate in our schools the benefits of sunshine and garments which will admit the health-giving rays to the body . . . if the work is carried out in an organized manner a nation of sunbathers may grow up as a result.

Well, Sir Bruce, I'm pleased to report, it worked. The war has been won. We take it on the beaches, in our gardens, in our lunch hours, anywhere and everywhere it appears. Male and female, young and old, have learnt how not to waste a single drop of its health – ahem, 'destroying'? – rays. In 1925 the Save the Children fund set up an artificial sunlight clinic in Hampstead, and the council considered the proposal that sunlamps should be installed in the public baths so that local children might benefit from regular basking.[20] Nowadays it is the young who are considered most at risk from such pernicious instruments, and are at the centre of the 'tanorexic' health scare.

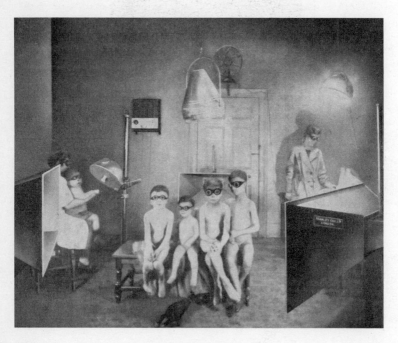

The young have always been susceptible to the fashionable allure of a suntan. Health policy on sunlamps, *c.* 1930

The wheel has turned full circle. The experts are now as evangelical in their opposition as they once were in their advocacy. But the propaganda worked too well. We're hooked. There were even signs of this emerging early on. In January 1932 the *Daily Herald* told how 'Drug-takers ask for sunshine', reporting how Londoners were queuing up at sunlight clinics for a few shillings' worth of artificial sunshine to get them through the winter. Tanorexics even then. From heliophobes to helioholics within about a decade.

SHE SHALL HAVE SUNSHINE

Outside gloom, fog and cold ; memories of a sunstarved summer; the threat of a hard winter.

Indoor sunshine and the delights of sunbathing. The thrill of health and the tingle of the blood in your veins, enjoyable relaxation in healthful ultra-violet rays at any time. The vitality and stamina that is proof against cold and winter ailments :—

home/un
SUNLAMP

Tanorexia in the making

Sunshine emerged as an idea or symbol, became a therapeutic agent, an hygienic panacea, a tonic you had to get your share of, and eventually a precious resource you desperately craved. There were some glorious summers and some abysmal ones – as there always have been. But new desires changed the weather, or our perception of it. As a writer in *Health & Efficiency* sighed during the poor summer of 1923:

since we cannot at will emigrate to a sunnier clime, we must hope for a more equitable allowance of sunshine during the remainder of the year; and . . . to utilize any gleam of bright weather that comes our way . . . Lucky children of Southern climes . . . lying in the sunshine and making a long, long meal of that inexpensive and abundant commodity!

This is a fair summing up of our national attitude to sunshine, and takes us back to our point of departure: the desire that impelled our East Ender made good – whose parents would have been the sickly slum-dwellers who caused the solar eugenicists such concern – to emigrate and luxuriate in this precious commodity. The solar revolution is complete. The sun-bronzed peasants of the south may be poor in material wealth, but are rich in sunshine. The gold is lying all around.

This is not the suntan as object of aspiration or fashion statement. This is full-on solar envy. It also sketches out the leisure travel blueprint of the post-war years, a new gold rush which exploited that free and abundant commodity, and turned the poorest regions of the south into veritable Eldorados for the sun-hungry north. But that's another story, another solar revolution we must now follow.

4

ESCAPE

> I do look forward to the sun and I go with a great store
> of . . . happiness that I can scarce keep bottled down
> in my weather beaten body.
>
> Robert Louis Stevenson, November 1873,
> on being 'ordered south' for his health

Baudelaire once compared life to a hospital, where all the patients are desperate to change beds. 'One would prefer to suffer near the stove, and another thinks he would soon recover near the window.'[1] Modern leisure travel best expresses this restlessness, and reveals some distinct trends and preferences. We in the northern wards mostly go southwards, where we know there is both a glowing stove and a vast blue open window. We travel to 'cure' ourselves of the stresses, frustrations and boredoms of our ordinary existence with the elixir of guaranteed sunshine.

Every day the skies above us are crossed and recrossed by people outsourcing this magic commodity. It is the principal reason for leisure travel from the UK, accounting in 2006 for nearly 70 per cent of its outbound airport traffic. It is so taken for granted that we don't even notice the heliocentricity of tour company names and logos. Sunshine is the only atmospheric phenomenon to be commodified in itself. People don't travel for snow, they travel for skiing or snowboarding. But we talk about 'getting' some sun, and many of the companies marketing products like 'winter sun' provide simply this: transporting sun-starved bodies from one hemisphere to another to consume it by the hour.

My own rather feeble survey at Stansted (the south-eastern springboard of the no-frills leisure travel industry) one spring weekend proved rather pointless. Asking why people were travelling merely elicited blank looks or versions of why do you bleedin' think, or mute pantomimic gestures towards the rain driving hard against the glass. This restless migration seems to be as instinctive as the birds obeying their call to the south each winter. We get the urge for going, and so we get up and go.

But this wasn't always so, and whilst the technological and economic developments that have caused an explosion in foreign travel over the last forty years might explain the volume, they did not create the desire which propels this relentless movement to the southern wards of the great hospital of modern life.

Although climate prompted some of the earliest and most significant movements of populations (to more fertile pastures or to escape floods, droughts or ice ages), weather is a very recent addition to the list of reasons for travel. If weather is mentioned in early travel accounts it was as something that got you there, or with which you struggled once you had arrived. Not an end in itself. You read of Crusaders boiling in their chain mail bags under the Palestinian sun, or poor ninnies expiring in the tropics because they failed to adopt appropriate clothing or habits.

When mad dogs or Englishmen went out in the height of the imperial sun, they usually paid the price. Edmund C. P. Hull, author of *The European in India*, a health manual for Anglo-Indians (1871), stated categorically that 'during the hot season it is unsafe for any European to walk out in the sun, *on the plains*, between eight o'clock in the morning, and four in the afternoon . . . without at least the protection of an umbrella'.[2] A seasoned Anglo-Indian himself, he recounts numerous cautionary tales of 'honourable and promising careers cut prematurely short' by the merciless tropical sun. Slip on a greatcoat, slap on a topee, and stick up an umbrella. Safe Sun, imperial style.

After plunder, pilgrimage, colonization, trade and mission, the next great motive in the history of travel was education. This reached its apogee in the eighteenth century, with the elite sending their young men to cultivate themselves among the scenes of classical antiquity. And when mind was forsaken for the body on the Grand Tour, then it was to the rites of Venus rather than Helios that these young classicists devoted themselves. References to the southern climate usually followed the approved Enlightenment belief that the sun was responsible for ensuring that these 'abject' peoples remained sunk in terminal torpor. The sun was seen as second only to the Pope in enslaving southern peoples, and the northern traveller could congratulate himself on originating from a climate that nurtured enterprise and freedom, as much as its opposite encouraged indolence and idolatry. If the sun had a place in the Enlightenment cosmology it was as a symbol of the reasoning mind. It was not something to subject oneself to so much as emulate, shining as it did into the dark recesses of folly and superstition. The Enlightenment sun was for light, not heat. Heat was suspect. When not encouraging idleness it promoted madness, fever and passionate excess.

It is not surprising that we find few sun-worshippers among early travellers, given what we know of the pre-twentieth-century prejudice against tanned skins, and the time it took for attitudes to change. It is unlikely that people would undergo the hardship of foreign travel just to bask in sunshine, if they had to be convinced of its benefits back at home, or be at death's door before they would be prepared to expose themselves to its rays. And yet, once again, it was exactly this that helped bring about another revolution in practices and beliefs. Health was responsible for the first weather tourism, as it was for the first suntans, encouraging the desperate to forsake their prejudices in milder, drier climes. Baudelaire's metaphor for modern restlessness is peculiarly apt and of its moment (1867). For here we see the northern ward of his hospital starting to stir.

O FOR A BEAKER OF THE WARM SOUTH

Lady Herbert's *Impressions of Spain*, also from 1867, opens with the following incontestable observation:

> What is it that we seek for, we Englishmen and Englishwomen, who, year by year, about the month of November, are seen crowding the Folkestone and Dover steam-boats, with the unmistakable 'Going abroad' look of travelling – bags, and wideawakes, and bundles of wraps, and alpaca gowns? I think it may be comprised in one word: – *sunshine*. This dear old land of ours, with all its luxuries, and all its comforts, and all its associations of home and people, still lacks one thing – and that is climate.

A costume drama version of a very familiar scene. And yet sunshine is a means rather than an end here; the ultimate object is health: 'For climate means health to one half of us', and without health 'the most perfect of homes . . . is spoiled and saddened'. People do still retire to the sun for their health, and 'winter sun' is still an attractive product for those who want to keep out the chills or cheat the season. But it's no substitute for the summer stuff the majority of travellers seek. The hotter the better. As Gal would say: 'We love it.'

But our ancestors didn't – not yet. Despite her effusive opening apostrophe, Lady Herbert scarcely mentions sunshine again over the next three hundred pages. When she does it is to complain about Seville in springtime, where the 'heat increases the fatigue' from seeing the sights.[3] Time to move on. Common sense dictated that most southern destinations were unbearable in the summer, even to the stoutest constitution. And so, when the first sufferers ventured south, they came in winter. To a large extent, they were simply exporting many of the beliefs and practices of the domestic coastal health resorts. Sun, sea and sand is now the established formula of the classic seaside holiday (from Bournemouth to Benidorm). But when the seaside was discovered and then ritualized by the English from the early eighteenth century, it was conspicuously lacking the first, to us, indispensable item.

When the quality took to the waters of Brighton or Scarborough to cure the spleen or follow the fashion, they took it cold. This was an essential part of the cure and the experience, one that was underpinned by the prejudicial climatics of the north. The beach was the site for invigoration rather than indolence, shocking the senses into vigorous mechanical responsiveness rather than caressing them into

relaxation and reverie. (Even Rollier, the pioneer of therapeutic sunbathing, stressed the efficacy of winter and Alpine
sun, and ensured that his pulmonary patients avoided anything but the chilliest air.) Southern beaches were viewed
with disfavour if not revulsion by those who promoted their
northern counterparts. They were described as boiling caul-
drons of mephitic, malarial danger and discomfort.[4] When
therapeutic necessity prompted a re-evaluation of such prejudices, this occurred reluctantly and cautiously. It was relative mildness the invalids sought, either returning to Britain
in the summer, or moving on to the spas or mountains of
France or Italy where cooling breezes offered welcome respite
from the scorching southern sun.

And they soon came in their thousands. In 1874, shortly
after the railway reached the Riviera, one contemporary
observer calculated 'that between seven and eight thousand
English invalids . . . annually spend the winter in the south in
pursuit of health'.[5] According to him, 'the great majority' of
these sojourners were consumptive. Resorts like Cannes,
Menton or San Remo were the last resort for many, clinging
desperately to lives that would inevitably terminate in foggy
England. The change-of-climate industry was big business. It
was hatched from a branch of medicine called Medical
Climatology which emerged from about the 1830s. This was
a specialist field, with experts detailing the density, humidity
and electricity of the airs, and the direction and velocity of the
prevailing winds at each resort, and prescribing to a nicety
which stretch of coast suited which ailment.

The invalid did well to consult such works as *Bradshaw's
Invalid's Companion to the Continent*. The Lonely Planet of
the valetudinarian sun-seeker, it offered tailored advice for
this very exacting traveller. And as opinion changed, or

advances in medical understanding were gained, so resorts or types of environment came in and out of fashion. If Rome drew such consumptives as Keats in the early part of the century, Robert Louis Stevenson and Aubrey Beardsley were joining the coughing caravan by coming to Menton after the mid-century; before the dry cold air of Alpine resorts such as Davos Platz started to rival, and then eclipse, the Riviera resorts at the century's end.

The health trade was a powerful authority and agent of change around the Mediterranean. The historian John Richard Green observed in 1876:

> Each winter resort brings home to us the power of the British doctor. It is he who rears pleasant towns at the foot of the Pyrenees, and lines the sunny coasts of the Riviera with villas that gleam white among the olive groves . . . At the first frosts of November the doctor marshals his wild geese for their winter flitting, and the long train steams off, grumbling but obedient, to the little Britains of the South.[6]

It's an evocative, but not inaccurate picture. As the British holidaymaker and émigré transformed the sleepy fishing villages of the Spanish Costas in the late twentieth century – turning them, as we lament, into little Britains, Blackpools in the sun – so the doctor put the Côte d'Azur on the map a century before. Nice, Cannes, Menton were completely transformed within decades of being discovered by British health-seekers and their advisers. The British physician was the resort developer or international travel company of his day, and those he sent hither start to resemble package tourists *avant la lettre*. If we laugh or sneer at the hordes of holidaymakers passively transported south like so many parcels today we should spare a thought for the delicate dispatches that set it all in motion.

Anticipating but not quite identical to the heliotropic hedonists of a hundred years later, the invalids came in search of relative mildness, not absolute temperature. The south merely made outdoor living possible during the winter months, and thus provided the conditions to effect a cure or delay a death. As Sir James Clark, one of the earliest exponents of the industry explained, the patient must 'consider the change as placing him in a more favourable situation for the operation of other remedies in the removal of his disease'. These included exercise, diet, rest and fresh air.

The medical experts were exacting on what was permitted those they ordered south. And although pleasure followed very closely on the heels of health in the Riviera resorts, 'making whole' was no holiday for our invalid sun-seekers. As Clark asserted: 'Every invalid who goes abroad, must make up his mind to submit to many sacrifices of his inclinations and pleasures, if he expects to improve his health by such a measure.' And so the doctors advised against attendance at dinners or social gatherings and excessive sightseeing. Exertion or excitement could be fatal. As could the throwing off of winter garments under the seductive influence of clear skies and gentle zephyrs. Dress as you would in winter at home, advised one. Never dispense with your flannels, stressed another, and always carry an extra wrap. Dr Bennet, a sufferer himself, claimed he never went anywhere without his trusty Inverness cape. It was still winter, and out of the direct sunlight it could be fatally chilly. So scurry home before sunset, and if you must admire the sublime grandeur of the setting sun, do it safely through the window of your villa or hotel.

And what of the sun – that which made the change-of-climate industry possible, and now the irresistible lodestone drawing millions of northerners along the routes this industry carved out? Most experts shared the common sense view that it was to be avoided. Clark claimed that 'One of the most exciting things to a sensitive invalid is exposure to a powerful sun.'[7] Too right. But for Clark this is a problem. Excitement is not on the agenda. And so the sun should be 'sedulously avoided, by resting during the middle of the day when the weather is oppressively hot'. Parasols (white and lined with green) were an essential part of the travelling kit for both men and women, as indispensable to the Riviera invalid in winter as they were to the Scarborough stroller in summer. Some invalids also sported blue or green sun goggles, to protect their eyes from the oppressive glare of the bright Mediterranean sunlight.

An essay from 1881 by Robert Louis Stevenson suggests the extent to which sunshine indulgence was anathema to the Riviera regime at the time. He had come to Davos Platz in Switzerland on his lifelong search for health that had earlier taken him to Menton. Coming north that winter, he remarks:

> To any one who should come from a southern sanatorium to the Alps, the row of sun-burned faces round the table would present the first surprise . . . The plump sunshine from above and its strong reverberation from below colour the skin like an Indian climate; the treatment, which consists mainly of the open air, exposes even the sickliest to tan, and a tableful of invalids comes, in a month or two, to resemble a tableful of hunters.[8]

In those days you went northwards to witness the bizarre spectacle of suntans. These tans were simply the by-product of the healthy exposure to the air in these Alpine sun traps, and not yet systemized into a cure in themselves. Stevenson takes an early documentary snapshot of what was quietly developing in the Alps, and which in time would allow sunburnt faces to be an acceptable feature of even Riviera tables. But not for another forty years. The northern prejudices were perhaps so deeply ingrained, the oppositions so orthodox, that solar exposure and its consequences could only happen, almost by accident, within the comfort zone of a northern latitude. Folk were just not ready to abandon themselves fully to the seductive southern sun.

SUNNY SPELLS

This was the general picture, but a few stray sunbeams were to be encountered among the sombre tomes I consulted. One in particular, Dr Bennet of Menton (he of the Inverness cape), dispensed advice of a decidedly modern character. Tubercular himself, he had come to Menton to find a pleasant corner in which to die. But he rallied under its genial influence, made Menton his winter quarters, and by publicizing its virtues, transformed it from a quiet village to a fashionable resort in the space of about fifteen years. He spent those years closely observing numerous cases, and concluded:

> Those who do best are those who accept their position cheerfully . . . and are content to lead a quiet, contemplative existence . . . if they can be satisfied to . . . [and] sit for hours basking in the sun, like an 'invalidated lizard on the wall', following implicitly the rules laid down for their guidance.

If the invalid could not join the giddy whirl of the pleasure-seekers, then there were perhaps unique compensations afforded this sedate tribe. And one of these might well have been sunbathing. Bennet describes a 'good plan for the invalid':

> [to] walk, ride, or drive to one of the many romantic regions in the neighbourhood . . . or, on calm days, to the pictur-esque rocky beach – to take the cushions out of the car-riage, if driving, with a cloak or two, and to remain sitting or lying in the sunshine, in some spot sheltered from the wind, for two or three hours . . .[9]

Now you're talking. It's sunbathing, but maybe not quite yet as we know it. It is winter, he is dressed for the Scottish Highlands. Perhaps his nose and eyes poke out from the thick muffler that surrounds his neck and the broad-brimmed hat shielding his delicate head. But he is lying in the sun, alone on a secluded beach that a century later will be packed to the gills with people doing the same, but wearing far far less. He is encouraged to bask, to give himself over to it, even to enjoy it. And so we glimpse a version of ourselves in (rough tweed) swaddling clothes, born amid salt foam and sun spangle on the shores of the Côte d'Azur.

Jost Krippendorf has identified the prevailing trends of modern travel. His *The Holiday Makers* (1987) concludes that the principal theme that 'runs like a thread through all these studies' is that 'travel is motivated by "going *away* from" rather than "going towards" something or somebody'. For modern travellers (or rather tourists), to 'shake off the everyday situation is much more important than the interest in visiting new places and people'.[10] If this is true, it is a relatively new trend. From plunder to pilgrimage, colonization to culture, the previous motivators have generally been incitements *towards*. Of course the big 'it' we all want to get away from in the north is the weather. In the Victorian invalid industry we see perhaps the first organized orientation away from the deficits of home; the first great escape. To quote Lady Herbert again: 'This dear old land of ours . . . still lacks one thing – and that is climate . . .'

These officially reluctant climate-hoppers are perhaps the forerunners of a shift in habits and sensibilities. For, as Krippendorf suggests, 'home' is far from perfect now. It means stress, frustration, routine, boredom – the epidemic syndrome that encourages us to self-administer the panacea of modern ills, and shift our hospital beds. Paid annual leave is a sacred right, a vital part of the social contract of our enforced labour, a safety valve on the late-capitalist pressure cooker. I'm run down, I'm working too hard, I'm getting stressed. I need a break. A few days doing nothing in the sun will sort me out. Now where's hot and affordable at this time of year?

The invalids showed us the way. Bennet's emphasis on the need for relaxation was shared by most authorities. It marked out the valetudinarian traveller from the tourist, the hectic flush of the consumptive from the hectic rush of the sight-seer. The able-bodied tourists were generally serious, studious and energetic: 'handed about,' as E. M. Forster puts it, 'like a parcel of goods from Venice to Florence, from Florence to Rome . . . quite unconscious of anything outside Baedeker, their one anxiety to get "done" or "through" and go on somewhere else'.[11] The invalid was exempt from such rigours and anxieties, and was explicitly encouraged to take things easy. Such indolence went against the official 'national temperament' of the northerner. Whilst the southerner might be considered languid, effeminate and lazy, northerners (and especially, so they believed, the English) had industry and enterprise in their blood and in their climate. Invalids learned to let go of these demands and give themselves over to the *dolce far niente*.

John Richard Green claimed that the doctor's advice to his charges in the south was 'the nearer you can approach to the condition of a vegetable the better for your chances of recovery'. Not only did he carve out the geography of modern sun-seeking leisure travel, the English physician also wrote its script. Relax, chill out, do practically nothing, advised the physician, and we willingly swallow the medicine first dispensed a century ago. Green describes the invalid's routine as a 'dawdle', involving 'a little music, a little reading of the quiet order, a little chat, a little letter-writing, and early to bed'. Apart from the last item, these are the established rites of the classic sun, sea and sand holiday. Trashy novels, personal music, a chat between dozes, and a few postcards dispatched home (weather lovely, hotel nice, waiters very friendly, hope it's raining back home . . .).

This emphasis also appears to be in marked contrast to the prevailing character of British seaside resorts at the time. From the mid-nineteenth century onwards, a holiday or day trip was devoted to the active pursuit of pleasure. As late as 1934, one of the first books to analyse the British seaside phenomenon remarked how 'Of all the ways in which people seek diversion, those of the seaside are amongst the least static . . . Movement is the order of the seaside day.'[12] They were cramming in diversion; escaping to a carnivalesque Utopia, allotted briefly to excess. Holiday-*making* was an active project. Peeling off, lying down, soaking up, simply *relaxing* in the sunshine – so natural and intuitive to us – must have been an alien concept. You had to be ill to indulge in it. And perhaps even to enjoy it, too.

Their days confined within a narrow scope, their concerns minutely focused on their health and its vicissitudes, our southern pioneers' thoughts turned incessantly to their senses. If modern sun-worship emerged out of medicine, the common denominator in all aspects of this story is the body – the site upon which sensations play and around which rituals are organized. The motivations and local details differed, but in the rhythms and rites of invalidism can be discerned the first prescriptions of our own cults of pleasure. Life for the invalid was not just vegetative, but 'simply barometrical', as Green put it: 'One's real interest lies in the sunshine, in the pleasure of having sunshine to-day, in the hope of having sunshine tomorrow.' They obsessively followed the sun, and so do we. The diseased, desperate body was the perfect instrument for effecting the revolution in sensibility that allowed sun-worship to take hold. Alive to the elements, alive because of the right elements, the invalid is an enforced sensualist. Despite the sternest strictures of the medical authorities, health and hedonism might have shaded into each other for those sedately basking in their own quiet spaces, away from the high roads of culture or the primrose paths of dissipation.

It is possible to detect a general thawing of northern frigidity in invalid accounts of southern exposure. Simple pleasures, unknown freedoms, were capable of blossoming uniquely at this latitude. John Richard Green's account of wintering in San Remo for his health describes this taking place:

> It is odd, when one is safely anchored in a winter refuge, to look back at the terrors and reluctance with which one first

faced the sentence of exile. Even if sunshine were the only gain of a winter flitting, it would still be hard to estimate the gain. The cold winds, the icy showers, the fogs we leave behind us, give perhaps a zest not wholly its own to Italian sunshine. But the abrupt plunge into a land of warmth and colour sends a strange shock of pleasure through every nerve. The flinging off of wraps and furs, the discarding of greatcoats, is like the beginning of a new life. It is not till we pass in this sharp, abrupt fashion from the November of one side the Alps to the November of the other that we get some notion of the way in which the actual range and freedom of life is cramped by the 'chill north-easters' . . . it is like a laughing defiance of established facts to lounge by the seashore in the hot sunglare of a January morning.

Who hasn't felt this rapture and release? A smug triumph at the disparity between what the calendar and our senses tell us. Tuberculosis may have compelled the author south, but this is soon forgotten as the fiery sun bath baptizes this once reluctant sojourner into new life. Pleasure, sensuous indulgence and a new sense of identity emerge out of the frail husk of the invalid. Ignoring medical orthodoxy, he throws off his wraps and dives headlong into the sense-bath of winter sunshine. It is still about relative difference – winter back there and winter here – but the excitement afforded by luxuriance in the southern sun is palpable. As is the sense of transformation. The greatcoats, furs and wraps are not just the uniform of the invalid, but the trappings of the upright, uptight northerner.

Green was a man of letters, but these rare pleasures are not the exclusive preserve of the sensitive intellectual. He even detects or imagines this thawing in the ordinary bourgeois. Under the influence of spring sunshine, 'the stolid, impassive English nature blooms into a life strangely unlike its own'.[13] The invalid is a sensual experimentalist, whose ailment has allowed him to stumble across a secret pleasure denied those whose robust health and hauteur steel them against such seductions. Throwing open the window on the northern hospital ward, letting this stimulant flood in to the self-imposed gloom of a heliophobic epoch, must have been an intoxicating experience indeed.

SOLITARY VICES

But we've reached the end of that epoch. A seductive new light is gathering and stealing through the blinds, as the twentieth century dawns. Change is in the air. The cult of invalidism (a fashionable fixation on one's ailments, and a cultivation of poetic pallor and sensitivity) was being replaced by the various cults of robust health, athleticism and the Great Outdoors encountered in our visit to the nudist camp in the previous chapter. Frail bodies, ravaged nerves were beginning to feel the elemental call of sunshine, and rejoice in the new freedoms it afforded.

André Gide made the kind of baptism Green described the theme of his first novel, *L'Immoraliste* (1902). Gide's semi-autobiographical novel is narrated by Michel who confesses to a life transformed under the influence of the southern climate. Michel is a scholar, who spent his early life in northern France, tranquil, cosseted and unaware of both the pleasures and perils of existence. His honeymoon in North Africa changes all this. Shortly after disembarking, he starts to spit blood – usually the death knell for sufferers both fictional and real, Michel's near-fatal illness actually marks the start of his new life. Passing through a pulmonary purgatory, he emerges into a new consciousness:

> I came to think it a very astonishing thing to be alive, that every day shone for me, an unhoped-for light. Before, thought I, I did not understand I was alive. The thrilling discovery of life was to be mine.

This light, of course, derives from the sun, the idol that draws Michel out of his former identity and initiates him into a new life of simplicity and sensualism: 'I see the sun; I see the shadow; I see the shadow moving; I have so little to think of that I watch it . . . Existing is occupation enough.' Not thinking is key to this scholar's solar rebirth. Oscar Wilde had once pointed out to Gide that the sun hates or deters thought.[14] Gide dramatizes this dictum. Michel's restless passage around the shores of the Mediterranean starts to take the form of an anti-Grand Tour. 'Deliberately disdainful' of his learning, he refuses to visit the approved sites. The south, as waves of invalids must have discovered, has something else to offer.

Michel's psychological transformation is accompanied by a desire to change his appearance. He resolves to strip away the 'secondary creature . . . whom education had painted on the surface' of his identity, and paint himself a new colour. Tan. Jealous of the 'beautiful, brown, sun-burned skins' of the Italian peasants, and 'shamed to tears' at the complete lack of colour of his own body, he disrobes in a secluded spot near Amalfi: 'I exposed my whole body to its flame. I sat down, lay down, turned myself about . . . Soon a delicious burning enveloped me; my whole being surged up into my skin.'

There we have it, probably the first account of deliberate sunbathing and tanning in a work of fiction. The year before Rollier opened his first solar clinic in the Alps, and Paul Zimmerman opened the first Freelight Park near Hamburg, Michel completes his cure through a form of heliotherapy. He feels different, so he wants to look different. The 'being' that surges up to his skin is his 'authentic' self, 'the "old Adam", whom everything about me – books, masters, parents, and myself had begun by attempting to suppress'.[15] Michel's tan is a form of cosmetic surgery, but impelled by primitivism rather than aspiration. To slough off class identity, to feel authentic, he needs to look like those who are close to the land. His real sickness has been civilization and its attendant repressions.

What is now deemed frivolous, vacuous and superficial – the acquisition of a tan – here carries profound symbolic significance. The sun 'cures' him of bourgeois identity and sexual inhibition. For, attending Michel's new identity and new image (he also shaves off his scholar's beard and lets his hair grow) is a growing interest in young boys. It is the restriction on loins rather than lungs of which he ultimately cures himself. Initiated in North Africa, it becomes a predominant focus of Michel's new life as a brown-limbed 'immoralist', and the semi-transparent, semi-autobiographical text of his confessions.

The sun allows new identities, true identities to emerge, and becomes an agent and symbol of sexual liberation.[16] E. M. Forster turned this idea into an organizing theme for his early fiction. Both *Where Angels Fear to Tread* (1905) and *A Room with a View* (1908) show buttoned up northerners finding varying degrees of freedom in response to the pagan promptings of the Italian climate. Where Forster trod cautiously – veiling his stories in nuance, irony and allegory – D. H. Lawrence rushed in with his short story 'Sun'. In many ways this is the heterosexual equivalent of Gide's novel, and also the first work of modern fiction dedicated exclusively to sunbathing. Published in 1926, it finds us once more in the decade when sun-worship really took off, and brings us to a significant turning point in our journey to discover its origins. Pleasure is about to rear its head.

The tale opens with a now familiar situation: '"Take her away, into the sun," the doctors said.' Juliet is a New Yorker, unhappily married to a grey, timid businessman. Her illness is never specified, but, given that this is Lawrence, it's probably a neurasthenic expression of her 'repression'. Initially 'sceptical of the sun', she allows herself to be packed off to Sicily for the winter. The doctor prescribes a version of heliotherapy rather than old-school change-of-climate theory, as he has advised Juliet to 'lie in the sun, without [her] clothes'. This she eventually tries, becomes a zealous convert to sunworship, and does precious little else for the rest of the story.

That is externally, of course. Internally a lot happens. Juliet is transformed by the sun: '"I am another being!" she said to herself, as she looked at her red-gold breast and thighs.' She becomes a fanatic, dismissing most of humanity, especially men, as 'so un-elemental, so unsunned . . . so like graveyard worms', and resolves that her own child will have his father's greyness sunned out of him. And so mother and child spend all day every day roasting in the now spring sunshine, growing more liberated and free with every deeper shade of brown. From frailty to rude health, in more than one sense.

'It was not just taking sun-baths,' Lawrence assures us, 'it was much more than that.' He can say that again. It being Lawrence, there are no prizes for guessing what Julia's sunbathing is *really* about. From the moment Julia first sees the 'naked sun stand up pure upon the sea line', the sun assumes a towering phallic status in Lawrence's one-track mindscape. The sun '*knew* her, in the cosmic carnal sense of the word' (whatever that means). The 'true Juliet was this dark flow from deep in her body to the sun'. Eventually she gets the hots for one of the local peasants, an elemental being like her new self, and she wishes he would be 'a procreative sunbath to her'.

Sun, sea and sex. The elements are all in place for a sizzling beachside romance. But this isn't sex in the sun (her lust for the peasant remains unconsummated). This is sex with the sun. And Lawrence lays it on with a trowel, giving a whole new meaning to the term 'sun stroke'. For in his hands a story about sunbathing becomes an onanistic incantation. It's enough to make even an obsessive like me turn quite queasy and wish he'd change the subject.

But, credit where it's due, Lawrence's tale does have an important place in the story we've been following. For, if we take his prose with a pinch of bromide, we do find in it the purest expression of the pleasure of sun exposure we have yet encountered. So over to David Herbert for the high-pitched equivalent of Gal's *basso profundo* that opened our song to the south:

> She lay with shut eyes, the colour of rosy flame through her lids . . . She could feel the sun penetrating even into her bones; nay, even further, even into her emotions and her thoughts . . . She was beginning to feel warm right through. Turning over, she let her shoulders dissolve in the sun, her loins, the backs of her thighs, even her heels. And she lay half stunned with wonder at the thing that was happening

to her. Her weary, chilled heart was melting, and, in melting, evaporating.[17]

Her wonder reminds us of how new the idea of what she is doing is, and how far and by what arduous ways we have travelled to find this simple act of relaxation and pleasure described. With this we can at last fully identify.

The sun has ridden high in the sky, but it is still carrying a heavy burden of meaning. We've had numerous explanations for sunbathing, and only now caught a glimpse of pure sensual pleasure in the experience. Can't taking sun baths just be taking sun baths? Can't the sun be just a sun, a tan be just a tan? Something people enjoy for its own sake, and the pleasure it brings. Not health, not identity politics or quasi-mystical sexual liberation? What of pleasure, leisure and vanity – sunshine's modern motivations? It's on the Continent in the 1920s where legend places the invention of the suntan. Surely someone's simply having fun in the sun. To find them we need to travel north from Sicily, to the French Riviera and the shore of a New World.

THE BEAUTIFUL AND THE TANNED

The French Riviera may not have invented the suntan, but it did help promote its fashionable allure. There is a good deal of myth attached to this also, and it's largely down to F. Scott Fitzgerald's novel *Tender is the Night* (1934). A character asks her hosts if they like 'this place'. '"They have to like it," said Abe North slowly. "They invented it."' 'It' was a way of life, lived outdoors on the beach and under the sweltering sun – in summer, so the story goes, for the very first time. 'They' are

PARIS - LYON - MEDITERRANEE

LA PLAGE DE CALVI.CORSE

Young, free and suntanned.
The new Bronze Age had arrived

Dick and Nicole Diver, a couple based in part on Fitzgerald's friends Gerald and Sara Murphy, to whom he dedicated his novel: '"This is only the second season that the hotel's been open in summer," Nicole explained. "We persuaded Gausse to keep on a cook and a garçon and a chasseur – it paid its way and this year it's doing even better."' 'Here' is a small corner of the French Riviera:

> about half-way between Marseilles and the Italian border, stood a large, proud rose-coloured hotel. Deferential palms cooled its flushed façade, and before it stretched a short

dazzling beach. Now it has become a summer resort of notable and fashionable people; in 1925 it was almost deserted after its English clientele went north in April.

Fitzgerald is immortalizing and embellishing events that did mark a significant change in attitudes and activities. In 1922 Cole Porter invited his old college friend Gerald Murphy and his wife Sara to stay with him at a chateau in Antibes for the summer. They came back the following year, staying at the Eden Roc hotel which was open that summer for the first time. They stuck around, taking on and transforming a villa which they renamed Villa America. Here they entertained numerous luminaries including Hemingway, Picasso, Dorothy Parker, Anita Loos and John Dos Passos. And, of course, the Fitzgeralds, who first came to the Riviera in the

Sunshine could make the east coast of Britain look like the south coast of France (graphically at least)

April of 1924. By the time Fitzgerald published his novel it had all changed. The Murphys left Antibes in 1932 when it was no longer the quiet exclusive Eden they had found it. By then the majority of hotels along the Riviera opened during the summer, and the season had been turned on its head.

Fitzgerald is the mythologizer of this transformation. His novel could even be read as an allegory of sun-worship in its journey southwards. It opens in a Swiss sanatorium, where the Divers meet. Dick is a psychiatrist, who marries one of his patients, the beautiful but fragile Nicole. They, like heliotherapy itself, come down from a northern sanatorium to the southern summer beach: 'We'll live near a warm beach where we can be brown and young together,' dreams Nicole. On the 'bright tan prayer rug of a beach', the Divers and the Murphys established the modern rites of sun-worship.[18]

The big breakthrough was the time of year. Whilst the invalid had generally been swaddled, shielded and apprehensive about giving him or herself over even to the winter sun, the new set revelled in it. Their day-time leisure routine revolved around the sun. But we know they didn't invent sun-bathing. Nor did they completely invent the summer season. By 1923 P&O were advertising summer holidays in southern France and further afield. An advert in *The Times* for that summer explains how 'Summer in the Mediterranean and Egypt has special and generally unsuspected attractions for holiday travellers':

> The atmospheric conditions throughout the middle months of the year are well nigh perfect. The daily sun-bath, between early tea and tiffin, away from the vagaries of our northern climate, is an experience which one will wish to repeat.

It proposed this as a 'new summer holiday scheme . . . within the compass of a modest holiday budget'.[19] The barbarians were already at the gates of paradise, looking for affordable summer novelties the very year the Murphys supposedly inaugurated the summer season, and Coco accidentally invented sunbathing. Yes, it is new and experimental. This ad is poised between two worlds, the final afterglow of empire – 'tiffin', I ask you – and the dawn of daring new freedoms. Try these unsuspected attractions. It's cheaper in summer, and really not as hot as you might fear. Here we detect a turning point in the meaning of summer.

The Riviera set gave high-fashionable status and *visibility* to a set of practices and ideas that we have been following along numerous disparate and relatively anonymous paths. The ingredients were all assembled: they turned it into a cocktail, and gave it a good Jazz Age shake. Apocryphal perhaps, but epochal none the less. They made modern sun-worship fashionable, and above all sexy.

Health isn't even mentioned. Nicole has been ill, but it is her mind, not her body, that needs rejuvenation. She tans because it is fun and young. A nonchalance and simplicity surrounds their attitude, wafting a welcome sea breeze across all the ardent intensity of what went before. They stripped it of its high seriousness and utility, and made it pure lifestyle. You are rich, young and leisured. Time is yours, the weather is yours. And for the first few seasons the beach was yours. Let the English scurry back to their Season and take their chances with their notorious climate. We'll stay here all summer long . . .[20]

The fashionable status of the tan helped give the solar revolution a final decisive impetus. From being conspicuously about one form of consumption (tuberculosis), the suntan became conspicuously about another. The fashionable tanner consumed sunshine, in abundance and at leisure, and conspicuously displayed the result for all to see. 'The 1929 girl must be tanned,' declared *Vogue* of that year; or, as the *New York Times* put it: 'for once the lords of fashion and learning agree, for doctors hail the present sun-tan fad as healthful . . .'[21]

The sun offered a new means of expression, the exposed skin a new creative canvas. The more you exposed at the beach, the more you could show to stunning effect later. The skin was now eloquent of where you sunned, how skilfully and for how long. By becoming glamorous the suntan definitively changed the social status and thus the value of sunshine. By making the effects of sunshine desirable it was a short step to commodifying it and marketing it as valuable in itself. The differential is, of course, time (ever weather's twin). By which I mean leisure. The deep all-over tan told of prolonged, deliberate and reliable exposure. Where have you been to get such a gorgeous tan? That's not two weeks snatched in the face of uncertainty at Blackpool or Southend. Too right it wasn't.

WELCOME TO THE PLEASURE DOME

We saw earlier how we in the north, and specifically the UK with its uncertain climate, got hooked on sunshine. How heatwaves taught us to binge, how poor summers and washouts caused us to lament, envy and dream as never before. If the opening scene of *Sexy Beast* shows the British working class consummating its love affair with sunshine – tying the knot with a lifelong commitment by emigrating to sunny Spain – the opening scene of *Passport to Pimlico* (1949) shows the courtship at the height of its adolescent infatuation. The film also opens under sweltering skies. But this is Britain, a Britain transformed by a heatwave. The camera comes to rest on another 'cockney' making the most of the weather. A young girl is sunbathing on her roof in her lunch hour. The radio wafts Latin exoticism across the scene, serenading her sun-soaked reveries with the strains of Len Norman and his Bethnal Green Bambinos. She wears sunglasses, a flouncy two-piece and foreign hair products. She is ripe for southern seduction.

But austerity still hovers over Britain even as the sun beats down. Rationing and the value of commodities determine their every thought and deed. But you can't ration sunshine. Discovering that Pimlico is actually Burgundy allows them to enjoy the exotic licence their new identity and the climate affords. 'Blimey, I'm a foreigner.' They enjoy relaxed opening hours, contraband delicacies, singing songs and staring at the moon long into the sultry night. A beguiling holiday romance bestowed by the weather. But like all holidays, and perhaps all romance, it can't last for ever. As Pimlico is restored to Britain so is the British weather. The film ends in torrential rain as it opened under torrid skies. Business as usual until the next heatwave imports the 'foreign' to this hemmed-in, ration-pinched, monochrome island.

But not for long. The film is poised on the brink of another epoch of increased sol-searching and sourcing. If the years following the First World War witnessed a new restlessness and discontent among intellectuals, who declared with one voice, 'I hate it here, and especially hate the weather, and so I'm buggering off abroad', the decade following the Second World War saw further aspirations and restless movements south – this time for everyone. The war had constructed air strips, and commissioned planes that could be put to good and profitable use. The theatre of war had given Tommy his walk-on part in some pretty exotic places, broadening his horizons and Technicoloring his dreams. Blackpool or Bognor would never look the same again.

Horizon was the name of the company that started this next solar revolution in 1950, the year after *Passport to Pimlico*. Starting with Corsican camping packages (two weeks for £32 10s. inclusive), it expanded to Spain, France, Sardinia and Portugal by the end of the decade. This was a revolution not of attitude this time, but of access. The desire was in place; only the means had been denied the majority of people. In 1951, 1.5 million people went abroad for their holiday; by 1965, the number had risen to 5 million. Paid holidays also increased over the next decades, with workers preferring

Beach pleasures. Brighton, 1930s

fewer hours to increased wages. Sun and relaxation had a value beyond material wealth.

John Walton, writing about the decline of the British seaside resort, charts the great holiday exodus from these shores. As he explains, by 1979, 'the number of foreign holidays taken by British customers passed the 10 million mark . . . By 1987 that figure had more than doubled, drawing in growing numbers of working-class visitors as the real cost of package holidays plummeted.'[22] The 80s and 90s witnessed the peak of this boom, with the majority of main holidays now taken abroad. If travel truly indicates a form of 'illness', as Baudelaire suggests, then here it has finally reached epidemic proportions. Some orderly obviously left a door ajar on the southern ward, and we've been streaming through ever since.

If southern France became the principal focus of the first wave of weather tourism, southern Spain was and remains the consummate creation of the second. Spain had not featured on the standard Grand Tour itinerary – too much of a detour on the royal road to Italy and Greece. Malaga had enjoyed a brief turn as a health resort in the first half of the nineteenth century, but for most travellers Spain was too far south, and so too hot. The first resorts to become fashionable in Spain (in the 1830s and 40s) were on the north Atlantic coast, and San Sebastián continued to attract even Sevillanas in summer up until the 1940s. When Laurie Lee first came to southern Spain in the 1930s he found the coast near Malaga 'a beautiful but exhausted shore, seemingly forgotten by the world . . . San Pedro, Estepona, Marbella and Fuengirola . . . were salt-fish villages, thin-ribbed, sea-hating, cursing their place in the sun. At that time one could have bought the whole coast for a shilling.' Surveying that same coast from the late 60s he claimed: 'Not emperors could buy it now.'[23]

The term Côte d'Azur (coined in 1887 by a poet) has an aesthetic appeal, conjuring up painterly visions of Matisse or Dufy. The southern Spanish coast seemed to know which side its bread was buttered, and the magic talisman that would transform its economy and its identity, when it became the Costa del Sol in 1930. From the late 1950s to the end of the century, 'the number of arrivals to Spain increased by over 4000 per cent',[24] the majority drawn to the beaches of the Mediterranean coast and the Balearics. By the early 70s, southern Spain had eclipsed the Riviera as the prime destination for northern European sun-seekers. A distinction it still holds today.

What transformed this coast was the transport infrastructure that could get volumes of holidaymakers there more quickly; the economies of scale that made the package deals and charter flights affordable for the masses; and, of course, the dependable hot sun that drew the millions southwards. A 1993 survey of 200 tourists in Torremolinos revealed that sunbathing was the most popular activity at 83.8 per cent, with sea-bathing next at 70.2 per cent. Only 27 per cent visited historic sites, and 15.7 per cent the countryside.[25] Sunshine was the vital ingredient in the new package holiday formula. It was what made foreign travel desirable, and what made the package model successful for those who exploited this precious commodity. A simple law of supply and demand took effect with a free but unevenly distributed resource. The holiday company got the tourists there, set up the pool and beachside infrastructure and left them to their own daytime devices: docile consumption of what they couldn't get at home. This was potentially a licence to mint gold, as sunshine moved from health into its leisure phase of product development.

The sun and sand formula had a value over and above the resources it supplied to an increasingly eager demand. What it added up to was a version of the full-rich life for vast volumes of people who, only a generation before, could barely dream of such things. In France, the Club Méditerranée holiday camps made this aspect even clearer. Starting in Mallorca in 1950, its tropical beach hut accommodation, banishment of radios and newspapers, turned any Mediterranean spot into a corner of paradise on the cheap. The sun was probably the only authentic ingredient.

The sun trade depended on economies of scale that made standardization unavoidable. The format established in Spain was reproduced ad infinitum and ad nauseam around the globe – wherever the sun shines, the sea is blue and the labour is cheap. The consequences are far from paradisiacal. Labour has been exploited and the landscape ravaged, natural resources are being used up at unsustainable levels, and local cultures, customs, values and languages are being eroded or eradicated by the monomaniacal desires of sun-seeking colonizers. Defenders of tourism claim that it has brought economic advantages to countless regions, who once 'cursed their place in the sun', and that it is patronizing and unrealistic for north-westerners to deny to others the standards of living they enjoy themselves. There are two sides to this argument, and no easy solutions to the desecration which the all-consuming tourist industry brings in its wake.

What is more easily agreed, however, is how we are supposed to view the typical beach holiday in a place like the Spanish Costas, whose mass-market 'paradise' is meant to be most right-thinking people's idea of hell.

In 1953, Robert Graves could be quite frank about why he came to live in Majorca in 1927: 'because its climate had the reputation of being better than any other in Europe . . . and I should be able to live there on a quarter of the income needed in England'. It also offered 'everything I wanted as a background to my work as a writer: sun, sea, mountains, spring-water, shady trees, no politics, and a few civilized luxuries such as electric light and a bus service to Palma'. Sounds idyllic, as Gertrude Stein, who recommended the island to Graves, asserted: 'if [you] liked Paradise, Majorca was Paradise.' Graves's priorities are hardly cerebral, and his emphasis on sunshine and cheap living no more exalted than those of the package tourists who were just over the horizon. And he a poet!

Twelve years later, Graves had cause to lament the very things that attracted him to Majorca. For by then 'a new idea derived from D. H. Lawrence's German-inspired sun-cult' had gripped the masses, and his paradise was invaded by the hordes, out for cheap sun. That year saw its millionth package tourist visiting Majorca. Graves feels compelled to point out that while 'May and October are Majorca's best months for the individual traveller . . . the fortnight-a-year vacationist can be served only in the summer'.[26] Time for the wheel to turn again.

Travel writing subscribes to a complete inversion of the tourist's hierarchy of values. When Paul Theroux got round to visiting the Med in the mid-90s, having taken many roads less travelled around the globe, he came in winter and studiously avoided the beaches and the sights: 'It was rainy and cool these October days. I became fond of this weather for various reasons. It was good for writing, and it kept the tourists away.'[27] 'Travel' is a relative term, and may be defined as 'not-tourism'. This is perhaps the only distinction left, as real travel becomes nearly impossible for most of us. The world has shrunk, aspirations have risen as air fares have fallen, and the travel game now starts to resemble the last frantic stages of musical chairs. It is only through conceit or denial that the would-be traveller can lord it over the bovine masses grazing in the ever-expanding pastures of the marketable world. Tourism is other people. As Paul Fussell puts it: 'it is difficult to be a snob and a tourist at the same time.'[28] Sunbathing makes this easier.

It allows the tourist to feel like a 'traveller', who need travel no further than Benidorm to assume this mantle of superiority. At least the touristic sightseer exhibits a spark of curiosity. Vain and vegetative, sunbathers look at nothing but the inside of their eyelids, other tanners, and their reflection in the mirror to check progress. If Gide and Lawrence hymned the sun because it hated thought and allowed feeling, others have found this a reason to condemn it as a vain, worthless pursuit.

The mass commodification of sunshine eventually spelt the relegation of the suntan back to its former lowly status. From being the brand of the peasant to the status symbol of the jet set, the conspicuous tan has now become the badge of the parvenu and the pleb. Both medicine and high fashion agree once again. Sunbathing is now vacuous, vulgar and dangerous. Strictly for chavs, WAGs and wannabes. All-over Burberry for the skin. When everybody has access to this stuff, the smart money, the old money, displays Olympian restraint – the mark of the truly refined when confronted with abundance. And so pale is interesting and ethereal once again.

AFTER SUN

But is sunbathing really irremediably soulless, witless and superficial? Putting aside the health issue, can't we trace some vestiges of seriousness and sacredness remaining? Unwrap even the tawdriest holiday package and there's surely a small flame of sanctity burning still. Sunbathing is deemed soulless because it is standardized and superficial. Give them sun, sea and sand and they don't care where they are. But by renouncing the aspirations, it also evades the delusions and disappointments of cultural or environmental authenticity. The rest of the world has been packaged, bastardized and consumed by travellers killing the thing they love (paradise publicized one year is a paradise lost the next). The body + sun = pleasure equation, however, remains pristine and untroubled by such desires and desecrations. The sunbather is at peace with his or her soul; is perhaps the last true romantic abroad.

Other folk arts are celebrated for their simplicity, reliance on natural ingredients, unchanging rituals and emotional engagement – be it cookery, viniculture, blues or sex. So why not the folk art of sunbathing? A late arrival in the travel parade, the sunbathing holiday is the least burdened with the trappings of touristic desire, delusion and materialism. Rush around in pursuit of authentic spectacles or experiences if you like. Find yourself in some obscure part of the Far East if you can. Or just take yourself to a beach closer to hand (leaving a smaller carbon footprint), and find the self that is always instantly there. Beneath the grey paving stones of drudgery and duty are the golden sands of authentic self-hood.

Sunbathing is back to nature, front to nature, and don't forget the tricky sides to nature too. It is the most innocent form of tourist ritual in that it asks little from the experience but itself. That the beach be a beach, that the sea be a sea and the sun be there, good and strong. The fact that you can reduce it to a formula, the unholy trinity of sun, sea and sand, testifies to its elemental simplicity. You don't need anything except the basics of environment and climate. A few libations, a towel or mat, but even these are extraneous to the fundamental experience in itself: the communion between man and sun. Surfing, for all its pretensions to sacredness, authenticity and hip, relies upon much more stuff, a pot more cash and a deal more hype to allow its communicants to break even.

Hell, you don't even need to bring a guidebook, you can trust in the packager to understand what you are looking for. Want last-minute sun? Click, pay and away. There is no risk of disappointment if you have relinquished any desires beyond the need to fly, flop and fry. There is no anxiety about being duped, arriving too late on the scene, or about being exploited or exploiting anyone else. On the unpretentious resorts of the Med, those particular horses have well and truly bolted. It is the same sun shining on Marbella as it is on the Maldives, or even Margate (if you're lucky). A good dollop of Hawaiian Tropic and you could dupe your senses into paradise.

Holidays don't last for ever; the sun isn't always here. By snatching this moment from time, the sunbathing vacationer experiences the quintessence of the holiday. A moment snatched from eternity, cradled under infinite blue, unadorned and prostrate before an ancient presence. For what could be more pure and ancient than the sun? Each 6ft by 2 of allotted sand becomes sanctified space, asserting the holiness of holiday through the making whole of mind and body: the body bronzed to pleasure soul. Under that brightening glance each one of us orientates (and then occidentates) towards our personal god. The sun puts the mass into tourism. Lawrence's heroine knew the sun in heaven, blue-molten with his white fire edges, throwing off fire. And though he shone on all the world, when she lay unclothed he focused on her. It was one of the wonders of the sun, he could shine on a million people and still be the radiant, splendid, unique sun, focused on her alone.[29]

If there are better grounds for worship I've yet to experience them. When Fitzgerald called the beach at Antibes a bright tan 'prayer mat', he wasn't wrong. The space dedicated to this devotion becomes hallowed. Purified and pristine in its ability to afford the unique relaxations, pleasures and recreations we need to cure us of modern life. Our flight from civilization so often takes us to the shore. And whilst the crowds prevent us gazing with solitary awe and melancholy yearning out to sea, we can close our eyes (we are forced to by the glare), shut out the noise of the shrieking teens with our headphones, and travel into ourselves, plumbing the depths of our identity in sybaritic solitude.

Soulless? Hardly. Baked meditative by that single fiery eye, sunbathing brings most of us as close as we'll ever get to the sacred. Your eyelids blaze saffron, the colour of sacred robes, and you ascend a few rungs on a ladder of gold.

O yes, Gal. We love it.

5

PLEASURE

> What wonder then if fields and regions here
> Breathe forth elixir pure, and rivers run
> Potable gold, when with one virtuous touch
> The arch-chemic sun so far from us remote
> Produces with terrestrial humour mixed
> Here in the dark so many precious things
> Of colour glorious and effect so rare.
>
> John Milton, *Paradise Lost*, Book 3, 556–62

Yes, but why do we love it?

I've probably given the impression that our taste for sunshine is something acquired relatively recently; a habit formed by fashion, economics and the theories of experts. But this can't be the whole picture.

Sunshine feels good. Surely we don't need doctors to tell us to get in to or out of it as scientific whim (or research grants) dictate. The cat knows nothing of aspiration, fashion or physiology, but instinctively seeks a strip of sunshine and purrs away the morning. It can't all be nurture. Nature must contribute something to this sensual banquet. But what part? Why does sunshine bring us pleasure?

You might assume that such an obvious enquiry would have been decided a long time ago. The ancient Greeks and Romans built solaria, and prescribed lying in the sun to cure melancholia.[1] So surely the reason for this had been pinned down to a neurochemical nicety years ago? Not a bit of it.

SMILE

Psychology seemed the obvious place to start. The diagrams in psychology books tend to resemble children's drawings rather than molecular monstrosities, so offer a relatively painless way in to the question. I found a handful of studies that approached the issue at a tangent. One detected a positive relationship between blue skies and higher stock market returns – sun out, mood up, markets up. Simple. Another equated grey Belgian skies with increased lottery expenditure (get me out of here, Lady Luck).[2] So much for effects; what about explaining the cause of such behaviour?

Some researchers at the Virginia Institute for Psychiatric and Behavioral Genetics at least tackled the question directly. They performed a series of experiments to explore how people felt and performed after exposure to bright Virginia spring weather. Better, strangely enough. But only when they were outdoors enjoying it. Those cooped up inside performed and felt worse than when it was grey outside. These researchers concluded that 'pleasant weather (higher temperature or barometric pressure) was related to higher mood, better memory, and "broadened" cognitive style during the spring as time spent outside increased'.[3] Now, I'm no scientist, but I can't help wondering about the circularity of this argument. I know what I find 'pleasant weather', but I thought scientists questioned such assumptions before they proved their hypotheses.

But maybe I'm being naive. For as the Virginian researchers assert, 'if lay psychology is to be believed, weather continues to be an important determinant of everyday mood and behavior in modern life. Given the pervasiveness of this belief, the paucity of scientific knowledge on how the weather affects human psychology is surprising.' Surprising, and also slightly disconcerting. If the white coats don't know, how are the likes of me to find any definitive answers?

The only way, I decide, is by conducting a little 'lay psychology' of my own. You needn't look at rain-sodden lottery tickets or buoyant stock market returns to prove that pleasant weather makes people feel more pleasant. There's evidence about us every day. Look up. What are the skies doing? Now look at the faces of the people you see about you. If it's sunny, chances are they'll be smiling; if it is cloudy, their faces will be cloudy too. Yes, I know that's a metaphor, but that's exactly the problem.

Talk about weather and emotion in the English language and you enter a metaphorical hall of mirrors. After the literal definition of 'The shining of the sun; direct sunlight uninterrupted by cloud', the *Oxford English Dictionary* lists several figurative meanings for sunshine, including 'a source of happiness or prosperity; a favourable or gracious influence; a condition or atmosphere of happiness or prosperity'. There are countless sayings and expressions that demonstrate this equation, and it would be pointless listing them. You'd need to be either blithely naive or pedantically obstinate to bother trying to demonstrate or dissect this relationship. But I feel I ought at least to try and accumulate some evidence of the connection.

Getty Images is the main image library used by the advertising and design industry. If you enter the term 'happy' into their search engine, it produces over three thousand pages with some sixty images per page. That's a lot of happiness. I can't pretend to have looked through all of them: after the first thirty pages I started feeling sick. It was like wading through candy floss. The healthy, toothy all-American grins began seriously to depress me. My jaw ached out of empathy. But I found what I was looking for.

Of the approximately 2,000 entries I surveyed, the majority made some reference to sunlight. Either the skies self-evidently symbolized the emotion (big, blue and sunny) on their own, or if these landscapes had figures they would mirror these conditions in their expressions. Faces radiated happiness; both bathed in and giving back sunshine. At least 65 per cent of the sample were shown outside in bright conditions. Or if inside, the sun would be streaming through. Of the sixty images per page, an average of forty-five to fifty were either in sunlight, about sunlight or used sunlight covertly to convey emotion.

Unhappiness yielded 250 pages. This was less clear cut, as unhappy people could appear outside, and the sun could be shining. But the majority of images were interior without conspicuous natural light. A few used rainy references – an unhappy woman with a rain-cloud symbol over her head; a woman staring morosely out of the window as the rain beats against the glass; two beach bums with surfboards frowning up at a lowering sky. The 'bad weather' in these images clearly communicated the emotion of unhappiness in itself, as the good weather had its opposite.

When I searched the term 'bad weather', Getty explicitly declared its preferences, and gave me the options of 'overcast' or 'stormy' to refine my search. Some familiar images from 'unhappiness' were even cross-referenced. If there is such a thing as emotional meteorology, it is recorded in the Getty image bank.

Getty serves the industries that sell dreams and the promise of happiness. It need not have any scientific scruples about meteorological bias. Not so the TV weather presenters, who are actively encouraged to be 'objective' in the way they deliver their forecasts. At the BBC, the presenters are scientists before they are personalities. Trained meteorologists on loan from the Met Office, they must present a precise and objective account of what the skies are likely to do, but in an engaging and lively manner. The BBC must entertain as well as inform.

It is the job of Andrew Lane, manager of the BBC's Weather Centre, to turn scientists into presentable presenters, and even 'personalities'. As he explained to me, 'No one wants their weather presented by Daleks. A forecast must be a conversation, with one imaginary individual in mind.' And so they are always unscripted. That was a revelation to me, but for Mr Lane it is essential. 'They have to be natural. If they were scripted, you might as well read it from the autocue yourself, or print it from the internet.' And so each forecaster holds the sum of complex meteorological data gathered by satellites and crunched by computers in his or her head, and tells the nation an extempore 'story' about what the weather will probably do.

Presenters must be able to think as scientists and engage as human beings. No mean feat. There are even guidelines advising against the use of value-laden terminology. Mr Lane occasionally has to take a forecaster to task for violating this code. Occasionally? For, despite being scientists, presenting objective facts for practical ends, the presenters quite evidently share our preferences. He'd need a Dalek to present a forecast for a May or August bank holiday who didn't betray some jubilation or disappointment at a prospective scorcher or washout. Predict three days of torrential rain without some conciliatory gesture or shred of hope, and you'd be lynched by a mob in knotted handkerchiefs and barbecue aprons. It's less about humanity than being emotionally in step with the nation: a nation tuned into this oracle of short-term happiness horizons.

But all I've done is itemize the obvious. What I really want to understand is the Science Behind These Smiles. That is the subtitle of Daniel Nettle's highly accessible and engrossing book called *Happiness*. There have been quite a few books about the psychology of happiness published recently (obviously a sexy subject), but none of them appears to be interested in the weather.

Nettle's *Happiness: The Science Behind Your Smile* does at least admit that weather might affect mood. But this is actually a variable to be discounted when seeking deeper insights. Nettle talks about levels of happiness. Level one is joy or pleasure: a transient mood that comes over us and is swayed by emotions. Level two is like the balance of satisfaction after a calculation of profit and loss. Level three is flourishing, the good life, fulfilling one's potential: such stuff as Utopias are made of.

Level one happiness can be influenced by something as trivial as finding a dime by a photocopier. Experimenters set this up, and asked people immediately afterwards about their satisfaction with life. As Nettle puts it, 'the participants sampled their emotional state, found the small joy of the dime still there, and so inferred that life was going pretty well'.[4]

This is mentioned in the same breath as an experiment by Norbert Schwarz, whose *Well-Being* is the standard textbook on the psychology of happiness. Schwarz phoned subjects on a sunny day, asked them to assess their sense of contentment, and found that most were doing just fine. If he drew attention to the weather Schwarz discovered that they were prepared to discount this factor and dig deeper into their emotional funds. Mood is merely the small change of the hedonic economy.[5]

Seeing that I was not getting very far on my own I contacted Daniel Nettle for some expert guidance in my quest. For him the lack of scientific interest in sunshine and happiness is understandable. He explained to me: 'Scientists who explore happiness are acutely conscious of the accusation of frivolity in what they do, and of spending time and money on investigating the bleeding obvious.' The sunshine–happiness equation pre-eminently falls into this category.

Nettle's Introduction almost makes this explicit when he talks about the struggles happiness underwent to be taken seriously by science. It has recently been rebranded as 'hedonics', as Latin – or Greek – always works a charm with scientists. For a long time such enquiry was dismissed as 'folk psychology', which was 'thought of by professionals as simply *bad psychology*'. This, he asserts, 'stood to psychological truth roughly as painting someone blue and doing a dance around them at sunrise stands to antibiotics'. But as Nettle explained, scientific research in this field depends on two things: what academics can get funding for (which in turn depends on the public utility of what they discover); and a need to provide answers. Because nothing can be done about the weather, there are few answers to be gained from it, and so no self-help books to be sold on the back of it.

So much for psychology. But there was something else both Nettle and I had independently noticed about recent works in hedonics: their cover designs. Nettle had even done an audit. Of the eight he looked at five either made conspicuous use of yellow, or had a blue sky in their designs. His own book was bright yellow, something he had objected to at first, but which he now feels was probably apt. Feels is the right word, because what psychologists are wary of exploring rationally inside, jacket designers are more than happy to exploit emotively on the outside. It would appear there is a colour scheme of happiness – the colours of a sunny day. His own reference to blue body paint and sunrise was perhaps not so random after all.

If this appears to be judging books by their covers, one of them, Richard Layard's *Happiness: Lessons from a New Science* (2005), goes even further. Layard is a professor at the London School of Economics, and the prime media pundit for hedonics. Dubbed the 'happiness Tsar' for his influence with the Labour government, his book is explicitly about providing answers. A cheeky spot-varnish sticker on the front of the paperback edition declares that the book reveals the 'Seven Causes of Happiness'. The publishers obviously have a keen eye for the self-help market. These answers do not include the weather, which Layard doesn't mention.

There is one reference to sunshine in Layard's book, however; just not a verbal one. If the content doesn't suggest it is one of the magic seven causes of happiness, the sticker advertising these insights clearly does. At the top of a blue spine sits a blazing yellow sun proclaiming the book's revelations. The hardback edition was pure blue with the word 'Happiness' curving into a bright yellow smile. Professor

Layard might not look to the sun for happiness, but his designers, marketeers and readers clearly do.

TRIP THE LIGHT FANTASTIC

The sun is second only to the yellow Smiley as a graphic shorthand for happiness in our culture.[6] I asked a sample of one hundred people to draw a symbol representing the idea of 'happiness' that was not a smiley face. Fifty-two per cent drew a sun. It was the only real contender; the next highest at 16 per cent was a rainbow. This isn't surprising, given that the Smiley is really only a rayless sun personified. You often

Mabli, aged 3. 'The smile makes it sunnier'

see the two combined. Give the sun a face, and his expression will radiate the emotion he both brings and symbolizes – a device used to sell anything from sliced bread to fruit juice and, obviously, sizzling summer holidays.

Children will invariably depict the sun in this way. (When

I asked my neighbour's little girl Mabli, aged three, why the sun she had drawn for me had a happy face, she told me: 'It made it sunnier.' Bless her. She also told me it's much better when there are two suns. Two suns? 'Oh yes, but I haven't seen that since I was little.') Copied behaviour or instinct? Maybe the latter, for childhood is that state when the doors of perception are thoroughly cleansed. And perhaps the only way to restore the vision splendid of childhood is to access your unconscious through meditation, hypnosis or by taking a pill. And not the ones that Mother gives you.

The Smiley is not unassociated within popular culture with certain controlled substances, namely LSD and Ecstasy. Both could be considered 'sunshine' drugs for a number of reasons. The imagery and references that surround them are saturated with sunshine; both had their cultural moments in sunbathed locations (California in the 60s, and Ibiza in the 80s); and both were the drugs of choice for two legendary hyped summers: 1967 and 1988, the so-called Summers of Love. On both occasions people were more than happy to paint themselves blue and dance around at sunrise. Here's why.

When Donovan called one of the first British psychedelic songs 'Sunshine Superman' (1966) he captured the mellow, yellow mood of an epoch. Sunshine is a psychedelic substance in this song, an atmospheric alternative to the trip the singer might have taken had he not 'changed his ways'. The song reached number two in the American charts in September 1966 just as the new locus of cool and creativity started to shift from London or Liverpool to California, and the sun assumed a momentous counter-cultural status. One of the first brands of LSD to hit the market was called 'Sunshine' (1967) (before then the drug was generally dispensed in liquid form, and the guarantor of quality was the chemist or lab from which it came). Variants were Orange Sunshine, Rose Sunshine and the infamous Blue Sunshine (star of the eponymous horror film from 1977 about its psychotic after-effects ten years later). The pioneers and propagandists of psychedelic drugs were mostly centred in California. It was here that Aldous Huxley first experimented with Mescaline in 1953, walked into his sunny garden and saw the world new made. And it was in California that the acid sun rose highest back in the heady days of the late 60s.

The psychedelic moment is a sunlit magic-lantern show of cultural memory. Song lyrics are bathed in solar reference, while 'radiant' best describes the psychedelic poster and record sleeve art from the period. As rainbow sunbursts

Solar flares. Hippies letting the sunshine in, Golden Gate Park, San Francisco, CA, 1969

announced the latest band or 'be-in' at the Fillmore or Avalon, the sun became a predominant graphic motif as of no time since the 1930s. 'Let the sunshine in' was a hippie greeting or rallying cry. Sunshine was an elemental expression of the hippie dream. Open your eyes, feed your head, bend your mind and fill your heart with nature's golden love. The hippie musical *Hair* (which opened in October 1967) ended that summer and the musical with the song 'Let the Sunshine in' as a final refrain.

Oh how we would if we could. But in Britain it was more wet coat than West Coast. The 60s were a very wet decade, conspicuous for a run of bad summers. The Move's song 'Flowers in the Rain' (August 1967) was probably closer to the mark.

We imagined our chance had come twenty years later in 1988, when the so-called 'second summer of love' ransacked the dressing-up box of hippie, and resurrected some of the imagery, associations and idealism of that jingle-jangle morning. The music was 'Acid', but the drug of choice was Ecstasy, the Smiley was its symbol, and its launch pad was Ibiza off the coast of sunny Spain.

Ibiza was something of a fallout zone for hippies when the world moved on elsewhere. The ethos of bohemian hedonism nurtured there found its fullest flowering in the decades that followed. By the mid-80s a few open air clubs had developed a distinctive style of music, energized by a drug that was relatively unknown outside the more outré gay clubs of Manhattan. The Ibiza moment coincided with the explosion in cheap summer sun holidays for young people. Some of the original British contingent who stumbled across the music and the drug there in the early 80s were just working-class kids on their first cheap trip to the sun. A group of young DJs brought both sound and substance back to London at the end of 1987, and attempted to transplant the sun-kissed vibe to wet London tarmac.

Danny Rampling opened up a club called Shoom near London Bridge (the name attempted to capture the euphoric rush E brings as you come up). It was Shoom that resurrected the Smiley, using it on a flyer for the club in January 1988, introducing 'the happy happy happy happy happy Shoom club, [as] Smileys bounced down the page like a shower of pills'. The symbol expressed the attitude of the club, which expressed the effect of the drug, which expressed the atmosphere of the island. Childlike, huggy, neo-hippies flooding the darkest days of Thatcher's Britain with a chemical surrogate for Ibiza sunshine.

MDMA, the chemical name for Ecstasy, belongs to a similar family to psychedelics like LSD, but demanded a new classification as an 'empathogen' (empathy-generating) drug. Before it was criminalized it was used by psychotherapists to help their patients open up their hearts. It might have been called 'empathy', but Ecstasy sells a greater promise to hedonists. And sell it did. Use of the drug and the popularity of the Ibiza sound (soon to be rebranded Acid House or Rave) exploded that summer. As an historian of the scene puts it:

> Ecstasy's innocuous appearance was the opposite of everything [young people] had ever been told about drugs . . . it came packaged, not as a drug cult, but as the ultimate entertainment concept, with its own music, clubs, dress code . . . Thousands of sunny smiles, the chatter of positivity, embracing total strangers . . . [7]

On Ibiza people still applaud sunrises and sunsets. It's part of the cult that has formed around a place, a chemical and a sound. The classic Ibiza backlist enshrines these associations, with anthems such as 'The Sun Rising', 'Sun is Shining' and 'I Am the Black Gold of the Sun' providing sonic souvenirs from the sun-kissed isle.[8] The first commercialized rave to exploit the new cult was called Sunrise, bussing clubbers out of London to trip about in the meadows at dawn. Whatever the weather, Ecstasy allows a chemical recommunion with that sunlit inspiration. A solar god transubstantiated into a small circular disc, and venerated in every ingestion. But there is a deeper reason for all these cultural references and associations. Whether your sun rises in Ibiza or Ipswich, pop that pill and you light a neurochemical path to your pineal gland. And that's why 'Everybody Loves the Sunshine'.

MDMA releases vast amounts of the neurotransmitter serotonin. This chemical is also naturally activated by sunlight. Here's the process, such as I understand it. Light enters the retina, and hits special receptors at the back of the eye. Some of the wavelengths are sorted into visual information, while others trigger chemical signals that travel along the suprachiasmatic pathway to the pineal gland located in the hypothalamus. It is here that serotonin (and its counterpart melatonin) are secreted into the bloodstream, and perform all kinds of vital functions in the body. Including the regulation of mood, sleep and energy. A deficiency of serotonin is believed to be behind Seasonal Affective Disorder (SAD).

SAD, only identified in 1984, generally affects people living at northern latitudes in winter. Whilst 1.4 per cent of the population of Florida are sufferers, this rises to 9.7 per cent in New Hampshire, 14.0 per cent in Oslo.[9] Research has suggested a direct correlation between the amount of bright sunlight received by the eye, and the levels of serotonin in the brain. Special light-boxes emitting bright broad spectrum light for winter mornings have been highly effective in treating the symptoms of SAD.

The eyes have been called the windows of the soul. This is principally for their revelatory function; but maybe this works both ways. Perhaps the soul is touched by light ('psychedelic' means soul-showing). SAD lamps give a serotonin boost, helping to lift the mood of sufferers so they can function normally; MDMA floods the bloodstream with the same chemical, lifting mood out of the roof and up with the rising sun. Serotonin is a sunshine chemical as Ecstasy is a sunshine drug.

But serotonin is not just a weekend hormone, it has a regular day job to do: helping to regulate the 'circadian rhythms' which all of us, and not just the ravers, dance to. The so-called 'body clock' is located in the pineal gland, and is believed to be operated by a rhythmic hormonal exchange. As a sunlight expert puts it: 'Sunlight helps us manufacture serotonin each day and this, in turn, provides us with melatonin. Also, sunlight switches off the catalyst in the brain that converts serotonin to melatonin. So the sun turns serotonin on and melatonin off.'[10] There are even SAD lamps branded 'Body Clocks' that work as alarm clocks, simulating the dawn rising with gently increasing bright light.

So here is something chemical, something of scientific substance to explain the metaphors and truisms behind our love of sunshine. Serotonin is the chemical link between the Smiley and the sun – the trippy icon paying intuitive homage to its natural source. But is this a full explanation for the riddle of sunshine and happiness? Hmmm, not quite. Would sunshine and unhappiness do?

There is a wealth of scientific material on sunlight and mood, but most of this is concerned with deficits and disorders. It is believed that up to half a million people in the UK suffer from SAD, and up to one in five of the population experience a milder form known as the 'winter blues'. It was the pathological consequences of winter light deprivation at northern latitudes that led Norman Rosenthal to identify SAD in the early 80s, which led to further research into the biological relationship between light and well-being. That need for answers again; a desperate need by those who suffer, and a responsibility in society to recognize those needs. SAD is a problem, so research funding is available to get sufferers back to normal, and back to work.

If a lack of sunshine makes some people depressed – and there is a probable chemical reason for this – then an abundance of sunshine makes most normal people happier, and for the same chemical reason. Surely? But I can only deduce this, as there seem to be no scientific papers explaining, let alone celebrating, why sunshine makes normal people even happier. The only research I found on 'normal' populations which took account of light was concerned with industrial productivity. It has been found that artificial lighting of a specific colour 'temperature' (the same wavelengths that are effective with SAD sufferers) enhances the productivity, customer service and 'well-being' of call-centre operatives in Stockport in winter.[11] Does 'well-being' also mean happiness, I asked Peter Mills, one of the authors of this study, whose company Vielife promotes environmental health solutions to businesses. 'I'd be reluctant to claim that,' he told me. 'It's not what we were testing . . . And it's not what most businesses are particularly concerned with.' There's no place for sunlit happiness in the Gradgrindian corporate environment.

Hedonics perhaps; hedonism, certainly not. You don't get funding for that. Unless you're discouraging people from enjoying sunshine. That's an entirely different story.

Recent scientific research suggests that
sunshine may have 'relaxing effects'

GOLDEN BROWN

Google the 'effects of UV radiation', and you find that
sunshine, instead of being relatively unimportant, is deadly
important. Evidence mounts and is duly trumpeted by the
media that the ultraviolet part of the spectrum brings skin
cancer, ageing, and probably a host of other nasties yet to be
discovered. There is no smiley face on the dermatological sun.

Every day they treat the (self-inflicted) wounds of solar
indulgence, and so it is not surprising that dermatology has
come to regard the sun as health enemy number one, and feels
a responsibility to re-educate a wilful public. The more they
discover, the worse the sun appears, the more absolute their
recommendations for avoidance. A paper on 'Sunscreens:
Practical Applications' in a textbook of photobiology recom-
mends that '*comprehensive* UV protection be used throughout

the day, every day, regardless of activity'. This is 'to protect the skin from daily UVR [Ultra Violet Radiation] assault in order to allow tissue repair of previous UVR-induced skin damage'. Another author in the same volume refers to ultraviolet wavelengths as 'outside aggressions'.[12] Dermatology considers us under 'assault' from a sinister enemy determined to do us harm. If the sun were a dictator, they'd have liberated his country by now.

Whilst the northern climate imposes a natural limit on how much of this dangerous stuff we can get access to, such restrictions do not apply to the sun's artificial agents. As names such as Midnight Sun and Endless Summer suggest, the indoor tanning salons can defy time, season and climate with instant, or even perpetual, gratification. As such, they have come in for the most virulent flak in the war on UV terror. A pronouncement from the American Academy of Dermatology in 2005 declared that 'because of the known health risk associated with UV exposure, the medical community has advocated that sunlamps should be banned for all but health purposes'. The British Medical Authority also pronounces its anathema for 'cosmetic purposes', and states: 'UV treatment should only be given under the supervision of a dermatologist.'[13]

These therapeutic dispensations (for conditions like psoriasis and eczema) invite comparisons with other controlled substances. Cannabis can be prescribed for certain conditions, and methadone is obviously available to help wean heroin users off their habit. UV radiation too begins to look like a narcotic in some of the more recent dermatological broadsides, and its devotees to be no better than junkies. In 2004 a paper was published by a research group from the Dermatological Unit at Wake Forest Baptist College in South Carolina, which provided a breakthrough in the scientific understanding of a possible 'addiction' to UV light. As they explain, most 'investigations into tanners' reasons for tanning have focused primarily on the perception of improved appearance'. But 'reported relaxing effects of tanning suggest the possibility of a physiologic effect of UV that drives tanning behavior'.[14]

Plenty of evidence supports these startling rumours, but this tends not to be reflected in the way tanning and sunbed use are represented in scientific literature. Vanity is usually seen as the prime motivator, and fashion the prime culprit. Such an emphasis both supports the need for effective counter-propaganda, and downplays the 'naturalness' of the activity or the desire to indulge in it (not to mention drawing a veil over medicine's own responsibility for the sunbathing fad in the first place). If biology rather than vanity drives this behaviour, then it is harder both to re-educate the public and to stigmatize those who indulge. Until now.

The Wake Forest group set up a clever experiment that blind-tested the preferences of 'frequent tanners' for particular sunbeds. Some of the beds emitted UV light as normal, whilst placebo beds were identical in appearance and heat output but did not emit UV light. The researchers randomized bed allocation for a few sessions, and then let the dedicated tanners choose which beds they preferred. The results were impressive, with a consistent blind preference for the UV lamps. Their bodies 'knew' what they liked. The researchers also monitored the subjects' reported moods after tanning, which also improved through UV stimulus. 'Relaxation and feelings of warmth' were how the tanners distinguished their preferred beds, and nothing to do with any improved 'pigmentation' resulting from the sessions. It was the experience, not their appearance that counted.

The possible explanation for this is found in certain neurological reactions between the skin and the brain. Until recently it was believed that natural 'opioids' such as beta-endorphin are only released by the brain. However it has been suggested that UV radiation can also trigger these feel-good chemicals through the skin. This is controversial and not quite conclusive as yet. But this theory does provide a plausible physiological explanation for the 'reinforcement' behind these tanners' habits.[15] Here at last I had found another scientific explanation for why sunshine brings pleasure.

Or so I thought, had these insights only been found in the pure temple of knowledge, and not in the sleazy back parlour of narcotics. The pleasure principle is immediately confiscated by the authorities. Endorphin is an internal morphine, having a similar function to the painkiller with a dual identity. Morphine in professional hands plays a vital role in alleviating suffering but, as heroin, reaps a harvest of dependency, crime and death. Endorphins are natural equivalents produced by the body. They are most famously associated with the so-called 'runner's high', achieved through extreme physical exertion. Athletes pushing their bodies to the limit deserve their little boost of neurochemical encouragement. Not so the tan-junkies luxuriating on their fluorescent opium divans. Delinquency is very close to vanity in the dermatological handbook.

Besides, not many generalizations can be made from these findings. For as the researchers point out, 'our sample consisted only of frequent tanners . . . as such we do not expect this group to be representative of all women in the general population, and we do not know if reinforcement would be found in less frequent tanners or non-tanners.'[16] These tanners might be a delinquent sub-species, whose physiologies are wired differently from the rest of humanity. Perhaps they are. The Wake Forest team have been doing some more experiments, and have recently published some more findings about UV 'addiction'.

They took a small sample of frequent and less frequent tanners. The explanation for their selection criteria shows how seriously they took the drug analogy: 'We included infrequent tanners rather than non-tanners so as not to expose indoor-tanning-naive individuals to conditions that we felt could potentially have addictive qualities.' They were no doubt aware of the legends circulating in drug lore of people being initiated in now-illegal recreational substances through medical or military trials. Wanting to avoid headlines such as 'Baptist Doctors Made Me a Sunbed Junkie', these guys were taking no chances.

In the new experiment the eight frequent and eight infrequent tanners were asked not to tan for two weeks. Having induced pale turkey, the researchers reached for a chemical called naltrexone, a 'narcotic antagonist that effectively blocks both central and peripheral opioid receptors'.[17] This drug is widely prescribed for heroin-users to wean them off their habits, by preventing the highs they crave. The researchers gave the blocker in various dosages to some subjects, and a placebo to others. They then randomized the UV exposure as before, and recorded preferences and symptoms. Two of the 'frequent tanners' pulled out of the experiment, complaining of nausea and jitteriness when exposed to UV light and with 15mg of this opioid-blocker in them. But with similar and higher doses, the infrequent tanners showed no negative side-effects from the blocker.

'Is it possible that frequent tanners experience acute or chronic elevation of opioid peptide levels,' the researchers wondered, 'and experience a form of opioid withdrawal in response to naltroxine?'[18] It looks like it. But, hold on, 'chronic', 'acute'? This is the language of pathology, applied to a physiological process responding to a 'natural' stimulus (UV light, albeit a surrogate in this experiment). Sunbathing has become an illness, or a form of delinquency, in the world of dermatology. If you like to sunbathe, I'd make the most of it if I were you. Our days in the sun may be numbered.

COVER UP

How could nature have got it so wrong? How could it make something so very bad for us also so very pleasurable? Given that the sun was invented first, and has an important role in the whole system, you'd think nature would have sorted out this design fault eons ago.[19] But it is a truism of psychology that nature never gets it wrong, and that pleasure was nature's reward for behaviour that furthers the species. The blessed invention of the orgasm is the most obvious example of this. But nutrition, exercise and other instances of 'adaptive behaviour' are also encouraged and rewarded by similar neurochemical caresses. The whole of sentient creation is supposedly manipulated in this way – all of us no better than rats in nature's laboratory, pushing the levers that to us say pleasure, but to Darwin say survival.

It's a marvellous system, and a great explanation. As evolution is the master key to so many human sciences, it also allows hedonics to be taken seriously by scientists. Psychologists call it 'adaption', when it's good, but 'reinforcement' when it's naughty. The pleasure of sunbathing is nature being naughty, and is apparently the great exception within the highly regulated pleasure circuit-board. A drug-related disorder, to be chemically blocked if we refuse to cover up.

No one tells lizards off for basking in the sun. Reptiles bask in sunlight for thermo-regulation, and to gain vitamin D.

Physignathus lesueurii recklessly laps up the sun in Brisbane, Australia

According to Jonathan Balcombe, an expert on animals and pleasure, they, like cats and humans, also do it because it feels nice.[20] Humans also need vitamin D. It was the discovery that sunlight could cure rickets (caused by a deficiency of the vitamin) that led to the enthusiastic championing of the tonic properties of sunlight by medical experts in the early twentieth century. This might be a plausible explanation for the pleasure factor, a euphoric reward for a very important chemical.

This can't, however, be stated categorically: the published work is either inconclusive, or biased towards the 'addiction' hypothesis. UV is in the dock, and guilty until proven innocent. *The Family Doctor Guide to Skin and Sunlight* sums up the case for the prosecution:

> Overall, it therefore seems that the UV radiation part of the spectrum may not be of any value to us at all, but instead is just responsible for most of the harmful effects associated with sun exposure, such as sunburn, photoageing and skin cancer.[21]

It's harder to imagine a more dramatic reversal of opinion in the history of medicine than the fortunes of the sun over the last fifty years. What was promoted as a medicine turns out to be a deadly poison. If I believed the new experts I'd have abandoned my enquiry long ago. As one declared: 'We now know that sunlight does not give happiness.'[22] Tell that to SAD sufferers, or to the inhabitants of northern Norway on the first day they see the sun again in spring, or even to Felix stretched out in the yard, blissfully unaware of how unhappy he is. But why should we believe one set of doctors telling us it's bad and to stay out of it, when another set seventy years before threw their equal authority into telling us it was good and to get into it?

There is, however, a growing counsel for UV's defence. In the US, a new sunlight crusade is being led by Dr Michael Holick of Boston University, a renegade dermatologist whose recent book *The UV Advantage* promotes the numerous benefits to be gained by moderate exposure to ultraviolet light.

For Holick, sunshine is a powerful medicine: 'Can you imagine what would happen if one of the drug companies came out with a single pill that reduced the risk of cancer, heart attack, stroke, osteoporosis, PMS, seasonal affective disorder, and various autoimmune disorders? . . . Well, guess what? Such a drug exists . . .' Dr Holick proffers plenty of evidence from his own researches and others' to support this claim for the marvels of sunlit vitamin D synthesis. Apart from the acknowledged bone and skin-related benefits, the most intriguing claim refers to the various internal cancers he believes can be prevented by sunshine.

For him, there is an urgent need for his message, as what he calls the powerful and influential 'sun-phobe' lobby have terrified Americans out of the sun, and into a silent epidemic of vitamin D deficiency. His figures put things in perspective: 'Fewer than half of 1 percent of people who develop non-melanoma skin cancer die.' This is the most common of the sun-related cancers bundled under the media scare-all of 'deadly skin disorders'. However, 'colon and breast cancers, which may be prevented by regular sun exposure, have mortality rates of 20 to 65 percent and kill 138,000 Americans annually.' As for the deadly but rare melanoma, 'there is no credible scientific evidence that moderate sun exposure causes melanomas'. The most significant counter-indication to support his scepticism is the fact that malignant melanomas tend to appear on places which are less regularly exposed to the sun.

Holick has an explanation for the great call to cover up: 'there are many billions of dollars to be made in emphasizing the only major medical downside of sun exposure (non-melanoma skin cancer) and not much money to be made in promoting the sun's many benefits.'[23] The sun is free, whilst the cosmetics and pharmaceutical industries he claims have vested interests in peddling fear and product.[24]

Richard Hobday is a UK sunlight champion. An architect and civil engineer by training, he is interested in all the benefits of natural light, and calls for the restoration of its importance in our daily life. His books, *The Healing Sun* and *The Light Revolution*, advocate a return to the lost art of heliotherapy. Sunlight is an holistic medicine in a very real sense, and Hobday details its numerous hygienic and therapeutic applications. It can kill bugs and bacteria if we consider it in the design of hospitals (demonstrated by Florence Nightingale, but forgotten by modern authorities); it can help lift depression in those suffering from mental illness (propounded by Hippocrates and recently confirmed by fresh studies); and it can cure and prevent the disorders identified by Holick. Hobday also points out how our refusal to acknowledge and utilize this powerful and ancient ally of the human race is behind a wealth of ills to which our urbanized, artificial and heliophobic society is heir.

The most fervent UK solar evangelist is perhaps Dr Oliver Gillie. Gillie calls sunlight a 'magic bullet', and demands a major change in the British government's policy on UV. Gillie claims that it is inappropriate for Britain to endorse a safe sun campaign that is based on the Australian response to its specific problem. As he argues:

> Northern Europe is not man's natural environment. Recent studies of human DNA tell us that man evolved in central Africa where the tropical sun provides plentiful ultra-violet light for vitamin D synthesis in the skin every day of the year. Lack of sunlight resulting from our northern location and our maritime climate makes the British Isles an extreme habitat compared with the tropical regions where human beings first evolved.

For Gillie, we need all we can get, and it is folly for the UK government to promote avoidance, which is leading to an increase in vitamin D related problems. 'Sunlight costs nothing and has very great health benefits. Sadly in the UK we get too little of it.'[25] His pamphlet *Sunlight Robbery* is effectively a sunbather's charter, encouraging the British public to get out into the sunshine at every opportunity. That's a medicine we are no doubt more than prepared to swallow; a return to the popular prescriptions of seventy years ago.

Two sides, opposing claims. Who are we to believe? I'm qualified neither to judge nor advise. It comes down to personal assessment of risk, how much you like the sun, and, at a macro level, the statistics. These are provided by some epidemiologists from Bristol University, who ask: 'Are we really dying for a tan?' They crunch the numbers, weigh up the odds, and conclude that the deaths from malignant melanoma directly attributable to sun exposure that could be prevented by the public health campaigns would be slight, and would not compensate for the risks presented by the chance of vitamin D deficiency. As they also, very sensibly, state:

> People find lying or sitting in the sun enjoyable and relaxing. This subjective sense of wellbeing may be important in itself in improving the quality of a person's life . . . Psychiatric illness is an important factor in population health, and any beneficial effect of increased exposure to sunlight might reduce appreciably the population burden of disease.[26]

If a boost of sunlight can make some unhappy people better, then it can prevent 'normal' people running the risk of depression or mental illness. Or just give them pleasure. At last.

EXPLORING THE PLEASURE SPECTRUM: EXPERIMENTS IN SOLAR ARTIFICE

Despite reading the scientific papers and consulting the experts, I still didn't have a definitive answer to the riddle of sunshine and pleasure. So, in one final push, I resolved to undertake my own research. I would use the sun's artificial surrogates in an attempt to isolate in a semi-scientific way the various components of the sunshine experience. I was prepared to be both experimenter and subject in this project, at one moment wearing the lab coat and holding the clipboard, the next slipping into my Speedos and prostrating myself on the experimental couch. Perhaps this was the only way I could tear aside the veil of indifference, prejudice and conspiracy, and bring forth the unalloyed truth from the stithy of empirical knowledge. Besides, it was winter and I needed a fix.

Experiment 1. The non-visible wavelengths of the electromagnetic spectrum

Though a fervent sun-worshipper, I had never used a sunbed or visited a salon. I am a purist, and, I must confess, was wary of their rather sleazy image. But having had my fill of dermatological defamation, I decided to make my own mind up about these much maligned solar surrogates.

Ciaran Mooney is the director of the Tanning Shop, whose company sought to overturn the industry's poor image. As he explained to me: 'twenty years ago, you would never see a salon on the high street, and few people would admit to visiting them.' They were hidden away, like back-street massage parlours. 'We put the sunbed in the windows, and made indoor tanning a respectable option.' For the public, perhaps; the health authorities of course have their own view. Mooney rises to the challenge, explaining that the Tanning Shop promotes 'responsible tanning', with strict regulations, and informed professionals to advise users and prevent misuse. The number of visits is restricted by a network-wide com-puter database which arguably regulates UV exposure, like an extra timer switch on the sun.

Mooney aims to provide an efficient, controlled alternative to sunshine for busy people without the time or resources to access the real thing. In winter it's a virtual holiday, and it was in winter that I made my first visit. He was right. There was nothing remotely sleazy about the bright blue and yellow (please note) salon I visited in Holborn, and no one I saw entering or leaving fitted the 'tanorexic' profile. Tanning lamps are about 95 per cent UVA, and 5 per cent UVB, with visible light and infrared. There were two types of machines to choose from: lie-down beds, and the new stand-up booths.

I tried both. Stand-up first. Clothes off, tan-accelerator cream on, and goggles fixed in place. In you step. Grab the bars, and on come the lights, heat-extraction fan and pumping dance music. I'm told most patrons dance. I attempted a desultory dad-at-the-disco shuffle, but it didn't feel right. I'm not usually naked and standing up when I sunbathe, and I wouldn't have chosen the music. But there was also something vital missing. It wasn't until I turned the fan down that I realized what it was, and made my first experimental finding. Heat is very important.

Infrared, which is responsible for the heat, plays no part in tanning, which is entirely derived from ultraviolet rays. But heat is an essential ingredient in making the sunless sunning experience more authentic. I mentioned this to the manager, who told me that patrons sometimes believe they have been short changed on their tans if the temperature didn't feel right. It is part of the illusion of simulated sunshine, but has no biochemical stimulant and no effect on pigmentation. This explains why sunshine still feels nice through glass, which cuts out UVB but not infrared. Heat is part of the feel-good bundle that is sunshine. Especially in winter.

The lie-down booth I enjoyed more. The horizontal inclination made it much more like the real thing. So I shut off the fan, turned down the music and toasted away the winter blues.

Experiment 2. The shorter wavelengths of the visual spectrum

Ultraviolet light is invisible. But it is only a nanometre away from the parts of the spectrum we can see, the first of which is violet. This colour occupies an important place in the symbolism and psychology of colour. Purple is the colour of power. The Romans institutionalized it as an emblem of status. At times only the emperor could wear it, his toga stained by the rarest dye extracted from sea snails found off the coast of Lebanon. To wear something so rare you must be very rich, very powerful, and therefore worthy of veneration.[27]

When Diocletian (245–312) became emperor he upgraded to golden robes and encouraged his court to adopt the Tyrian purple. The emperor was an avid devotee of the cult of Sol Invictus (the unconquered sun), which was the official state religion in the empire's final pre-Christian days. The symbolic importance of gold and purple survived when the empire adopted Christianity as the state religion under Constantine (another former sun-worshipper) in 325. The golden halos adorning sacred figures are the most obvious solar legacies from this transference of symbolic assets. But the imperial purple was retained by the priests, Christ's elect emissaries on earth, whose dress bears evidence of this sartorial signature of the Roman elite to this day.

But the papal penchant for purple owes as much perhaps

to photobiology as the exquisite pomp of ecclesiastical drag. It expresses *lux*ury in its purest and most literal sense. As purple resides in the violet section of the spectrum, it is at the very borders of the visible and the invisible, the mundane and the mystical. Purple is the hardest colour for the human eye to discern. It is only just visible, quivering iridescently between two realms: *ultra*-violet – the powerful, mystical presence beyond; unseen, but keenly felt by worshippers as an ecstatic radiant fire. Small wonder its visible equivalent was adopted by sun-worshipping pagans, and continued in the bright raiment (ray-ment) of their priestly successors.

These speculatious were somewhat fanciful, so I consulted an expert on colour. Angela Wright is the founder of Colour Effects, and a leading exponent of Colour Psychology. But if invisible wavelengths can be allowed to have profound effects, why not the non-chemical, visible ones? Why not indeed, agreed Angela. What she told me made a lot of sense. Colour is all about emotion, with different colours producing different emotional effects. And as colour is entirely dependent on light, sunlight triggers a more intense emotional response. As Angela puts it: 'colour is light, and light is the source of life.' It turns up the volume on life.

Colour psychology and colour symbolism are not the same thing. Colours mean different things in different cultures. But sometimes symbolism and psychology agree, as they do in the colours I'm interested in – purple and blue. Angela explained that 'the shorter the wavelength, the more profound the psychological effect.' Violet has the shortest wavelengths within the visible spectrum. Symbolically it is the colour of spiritualism, and the higher mind.

She also pointed out that in Eastern thought the seven chakras (the body's centres of energy which govern the mind–body relationship) are allocated colours, with the seventh and highest being white or violet. Interestingly, these seven chakras move down the spectrum from white/violet all the way to red. Indigo, which resides between violet and blue on the spectrum, is allocated to the sixth chakra, and identified with the pineal gland (the so-called 'third eye'). Its elements are light and time – the very function with which this gland is associated in Western scientific thought, responsible as it is for regulating the body's 'internal clock'. Blue, the next colour in the visible spectrum, is also important for both symbolism and science.

Blue, like purple, has a special status in what might be called the chromatics of power. It was the exclusive colour for depicting the Virgin's robes, and, as ultramarine, it was the most precious and expensive pigment for painting, illumination and stained glass windows. It is still associated with power and success in the West (from blue blood to blue chip), and is the favourite colour for use on companies' annual reports. According to Angela, blue is also the world's favourite colour.

If you recall, serotonin (which we may dub simplistically the sunshine hormone) is believed to be triggered by light stimulating special receptors at the back of the eye. Recent research has indicated that the photoreceptive cells that have this chemical effect are neither the rods nor cones that carry information from the other parts of the visible spectrum, but a third type of cell (termed 'Melanopsin receptors') which appears to react most effectively to a specific region of the visible spectrum. This region is, of course, blue. This was the part of the spectrum used in the lights that enhanced the productivity and well-being of those Stockport call-centre operatives. It is also the optimum wavelength range for treating

SAD. Blue is the colour of 'well-being', then. Scientists appear to have just discovered what designers of books about happiness knew or felt all along.

I was first alerted to these findings by Steve Hayes, founder of Outside In, the UK market leaders in designing the special light-boxes used by SAD sufferers. Steve is something of an authority on the subject, who very kindly shared his insights, loaded me with research papers, and sent me back from Cambridge with one of his special light-boxes to use in my next experiment.

I had also never used a SAD lamp before. Despite my obsession with sunshine, I don't actually suffer from SAD. Even so, I thought, more sunlight in winter certainly couldn't hurt. So I fixed the light over my desk, switched it on, and forgot about it while I worked. After about a week of using it each morning I discerned an effect. But not a positive one.

I'd got so used to the artificial sunlight inside that I could hardly face going out. I don't think you can get hooked on serotonin (even Ecstasy isn't physically addictive); but I appeared to be suffering depressive withdrawal symptoms when forced to prise myself away from my lamp – the opposite of what these lamps were supposed to achieve. So I limited my usage to the suggested ninety minutes, and persevered with my research.

The light emitted from the rectangular box fixed above my writing desk was certainly bright, and had a bluish tinge to it. Whilst this is, of course, biologically right, it's aesthetically wrong. Sunshine and the sun have 'warm' colours in our conditioning, and so, like the lack of heat in my tanning bed experiment, something didn't feel quite right.

One morning in February I was able to make a direct comparison. I'd been using the lamp all morning, and then at about 11 o'clock the real sun came out. So I shifted my laptop over to the window through which the sun now streamed. I immediately realized how the brilliant orange colour that lit up my closed eyelids had been lacking from my lamp. I had to put my face up close to the lamp before I got the same effect. Not a very comfortable or productive position, so I returned to the window and basked. The clinical blue of the light-box seemed peculiarly anaemic by comparison. It was simply not the same.

But then, it doesn't pretend to be. It is not there to make me happy, but to make SAD sufferers better. And I have never suffered from SAD; in fact, the opposite. Whilst sufferers tend to get depressed at the onset of winter, I often find myself depressed in 'summer' – frustrated and even right-eously indignant at the failure of repeated summers to deliver on my expectations. I generally feel a degree of relief and res-ignation in autumn once the hope of some mythical ideal has passed. It's the principle of the thing. My solar fixation appears to be as much nurtured as natured, associative as well as physiological.

Which led me to my third and final experiment. This wasn't an experiment at all, but a public demonstration of the sublime poetry of sunshine, and not of my instigation.

Experiment 3. The longer wavelengths of the electromagnetic spectrum, some mirrors, misted water, a converted power station and thousands of members of the general public

In the winter of 2003–4 Olafur Eliasson's installation *The weather project* drew one of the largest crowds London's Tate Modern art gallery has ever seen. A vast artificial sun flooded the Turbine Hall with an eerie golden light in which the

Olafur Eliasson's *The weather project* banished the London winter in the Turbine Hall at Tate Modern for the Unilever Series, 2003

nation flocked to bask. But it was literally a trick of smoke and mirrors. Hundreds of sodium yellow bulbs (the same used for street lights) formed a semicircle behind a translucent screen. The sphere was completed by a mirror that spanned the entire ceiling, creating soaring infinity for the artificial sunscape, and reducing the spectators' reflections to tiny matchstick figures high above our heads. A fine mist created atmospheric texture to complete the climatic transformation of the former industrial space.

The weather project was a vast statement in solar sublimity – combining a Turner canvas with a great rose window in this cavernous cathedral of art. An elementary rose window, now purged of the picture-book superfluities used to awe the unlettered into dutiful devotion in Chartres or Saint-Denis. Coloured lights and pictures we have a-plenty these days, but the sun still exerts its primal pull, and the sun we can't switch on. But that's just what Eliasson did.

High spirits buzzed around the Tate hall and there was a real holiday atmosphere (not something usually encountered at art galleries). Picnics were taken. Groups lay on the floor and formed kaleidoscopic patterns in the mirrors overhead. Those who didn't frolic basked instead in solitary contemplation. In other words, the public treated the Tate like any beach or public park on a sunny summer day. Some spectators simply stared. Not with the quizzical or studious incomprehension that often characterizes our encounter with modern art, but with faraway primal recognition. Resembling those vast shimmering balls of fire beloved of films set in Africa or Australia, this sun struck a chord of atavistic adoration buried deep inside us. We may not know much about art or the weather, but we know what we like.

Eliasson's *weather project* is entirely symbolic and associative. A project in a variety of senses, it absorbed and radiated the projected desires and memories of a nation of sun-worshippers. We completed the artist's work, as the mirror completed his illusion. Of all the artificial suns I've tested, his is perhaps the most faithful and potent: the closest to the sun we carry in our hearts and in our heads.

If they needed an industrial scale placebo for SAD lamps they had it here. Reaching souls rather than synapses, Eliasson's yellow bulbs provided no warmth, no ultraviolet rays; they were neither bright enough nor the right colour frequency to stimulate the production of serotonin. And yet, fooled like birds in a cage, we sang to his sunshine.

6

LOVE

Busie old foole, unruly Sunne,
   Why dost thou thus,
Through windows, and through curtaines call on us?
Must to thy motions lovers seasons run?
                    John Donne, 'The Sunne Rising'

STORMY WEATHER

I was once asked my view on the colour of love. I decided it was the colour of the sky. Deep blue and infinite when new; bruised, turbulent and menacing when passions or tempers are at their height; grey, monotonous and without horizon when love has fled.

A predictable enough response, I suppose, given my obsession. But such thinking is not without precedent, and might be considered a prime example of what the Victorian critic John Ruskin termed the 'pathetic fallacy' (1856). This is a defining characteristic of much poetry and painting, but Ruskin disapproved of it. He complained how 'violent feelings', by which he meant strong emotions, coloured poets' depictions of the natural world.

Ruskin considered this a 'fallacy' because he believed that nature should be rendered as it actually was, and not sullied by the desires and despairs of fallen mankind. For him, temperaments that displayed this tendency were those 'borne away, or over-clouded, or over-dazzled by emotion'.[1] His metaphors are revealing. Whilst his term isn't exclusively applied to the weather, this is one of the most common recipients or carriers of emotion in art. If in the last chapter I sought to understand why certain types of weather affected our emotions, here I'm interested in the reverse: how our emotions affect the weather, or at least our perception of it.

We talk of people having certain 'temperaments'; of there being a certain 'air' about them; of their being 'cold', of a 'sunny disposition' or permanently under a cloud. A temperamental person is one who changes like, or perhaps even with, the weather. If humans are subject to all the 'skiey influences', as Shakespeare put it, if their own moods can be compared with those skies and their incessant shifts and caprices, it's not surprising that we often attribute moods to the skies themselves. How we feel often determines whether we see the heavens as a majestical roof fretted with golden fire, or (as Hamlet does) 'a foul and pestilent congregation of vapours'. There is so much overlap between the emotional and the meteorological spheres, and the fabric separating them is so permeable, there is little point in bemoaning their creative commingling. As Ruskin's own works clearly show.

In 'The Storm Cloud of the Nineteenth Century' (1884) Ruskin shows how weather can be the perfect vehicle for strong emotions. He claims to have discerned a decided change in what he refers to as the 'character' of the weather over recent years. As he claims, in the old days:

when weather was fine, it was luxuriously fine; when it was bad – it was often abominably bad, but it had its fit of temper and had done with it – it didn't sulk for three months without letting you see the sun, – nor send you one cyclone inside out, every Saturday afternoon, and another outside in, every Monday morning.

We have recently come to accept that the state of our own atmosphere is an urgent political issue, and take seriously the view that human causes can have dire meteorological effects. This was Ruskin's conclusion over a hundred years ago. He blamed the 'Storm Clouds' and 'Plague Winds' of that time largely on the factory chimneys he saw belching out pestilential pollution all around him. As such, Ruskin has been hailed as the 'first green', a prescient eco-worrier who witnessed the sowing of the industrial harvest we are now reaping.

At the time, however, many saw his tirade as so much hot air, and further confirmation of the old sage's descent into madness. But Ruskin's tirade didn't point to mental illness so much as to a very human tendency to imagine a sympathy between our own emotions and the world around us. Ruskin was getting old, and so was the world, and, of course, the weather was better when he was young (as it was for everyone). And here his own 'pathetic fallacy' comes to his rescue. There is certainly pathos in his diary entry of July 1871 when he first noted the change:

> through meagre March, through changelessly sullen April, through despondent May, and darkened June, morning after morning has come grey-shrouded thus.
>
> And it is a new thing to me, and a very dreadful one. I am fifty years old, and more; and since I was five, have

gleaned the best hours of my life in the sun of spring and
summer mornings; and I never saw such as these, till now.[2]

Such despairs are but a heartbeat away from imagining nature
to be in tune with our feelings. If it is a fallacy to respond in
this way, then it is an understandable and highly prevalent one.

The weather doesn't 'sulk', have fits of temper or feel
despondent. But humans do. Give them a public podium, a
spell of bad weather, and they can out-Lear Lear in calling on
the heavens to bear witness to their despair. In fact Lear's performance in Shakespeare's play is a perfect example of why the
pathetic fallacy shouldn't be considered a problem. It is a widespread and deep-seated response to the universe, and occurs
repeatedly in some of the greatest works of art. From John
Milton to Thomas Hardy, from Geoffrey Chaucer to Dylan
Thomas, indeed, to Bob Dylan, poets have found the weather
a peculiarly powerful and appropriate metaphor and mirror for
the strongest human emotions.[3] And for good reason.

Weather is, always has been and always will be, a problem.
From the moment figures emerged on the landscape they
were obliged to find ways to cope with what the heavens
unleashed. Meteorology is an attempt to impose order on or
find pattern in chaos. It is literally the science of meteors –
things that fall from on high – and its official history records
the heroic struggles of some of the greatest minds to explain
and predict these things. With mixed results.

It is in its predictive capacity that we now chiefly encounter
meteorology. From the turn of the nineteenth century, the
observation, collection, interpretation and dissemination of
weather data for prognostic purposes has become increasingly
systematic, coordinated and professional with every decade.
Modern meteorology sends satellites into space; sets up

recording equipment in some of the most inhospitable corners of the world; has achieved a quite extraordinary degree of cooperation across countries, where the need to collect reliable real-time data has overridden otherwise insurmountable differences between nations; it employs some of the largest, fastest computers in the world to make nearly infinite calculations with this data; and assembles, interprets and presents the most likely prognoses for the weather days, and sometimes weeks, in advance. The Met Office can now boast that, thanks to these innovations, its three-day forecast 'is more accurate than a one-day forecast in 1980'. Its objective is constantly to improve the accuracy of forecasting, and its main message to the world is 'we're getting there'.

And yet, as meteorologists are ready to admit, even with all this kit producing all that data, things can still go wrong. Any number of factors can get betwixt cup and lip in forecasting. Fronts unexpectedly collide, hurricanes change their minds and wreak havoc with an unexpected landfall, and clouds and rain can spoil what promised to be the perfect weather for the perfect June wedding. Now isn't that ironic? Weather must surely be the most baffling, tricksy and infuriating realm of experience to have tasked the intellect, patience or reason of man, from the dawn of time. After love, of course.

When Eve persuaded Adam to take a bite of that apple, all kinds of nasties entered the idyllic world they had known until then. With the Fall, came sin, shame, labour, pain and marital conflict. But it also introduced Death, Time, the Seasons and the Weather. They are effectively the same thing. As John Milton puts it, before Paradise was Lost, 'spring / Perpetual smiled on earth with vernant Flowers, / Equal in Days and Nights . . .' But as punishment for Adam and Eve's transgression, God introduced the seasons. He did this by fixing the sun.

> The sun
> Had first his precept so to move, so shine,
> As might affect the earth with cold and heat
> Scarce tolerable, and from the north to call
> Decrepit winter, from the south to bring
> Solstitial summer's heat . . .

The result was all kinds of 'noxious' weather. Adam, 'hid in gloomiest shade', sees all this unfolding, and laments: 'O fleeting joys / Of Paradise, dear bought with lasting woes!'[4] As well he might, and us with him. For this, alas, is the reality of the fallen world. Life now had a duration, winter followed autumn, and Adam and Eve had to clothe themselves. Not just out of shame at their nakedness, but because it was no longer perpetual spring, and it had no doubt got a bit parky.

Time, the Seasons and the Weather, therefore, derive from that first moment of disobedience, and the first pains and problems of love. They've accompanied love on its troubled course ever since. Switch on the radio, listen to pop songs, and pretty soon the singer will be making this connection. It's so prevalent it tends to go unnoticed. But once you're tuned in you'll find it everywhere. Like the weather itself. This is hardly surprising, for it draws on the weather's usefulness as a peculiarly expressive form of emotional shorthand. Pop songs are three-minute emotional symphonies, and so those about love are often awash with weather.

THE SUNSHINE OF YOUR LOVE

We know that sunshine can make people happy, and is a prime metaphor for happiness. Happy love therefore has a fairly predictable colour scheme in pop. Katrina and the Wave's feelgood anthem 'Walking on Sunshine' (1985) borrows the emotional charge of sunlight to express ecstatic love at its bursting blissful height. Stomping beat, bouncy brass, as shiny and yellow as the sun itself. It's instant aural sunshine, and obligatory airplay at the merest peep of spring. I don't even need to quote from it, as you're probably singing it already. The song's entire sentiment can be summarized as I'm in love and it feels good. It feels good as sunshine feels good. And that's about the best way I can express it. It's not clear whether the sunshine is real or metaphorical in the song. It may even be winter outside, but in Miss Katrina's heart, in her step and in her song, it's glorious sunny spring.

Love is very like sunshine. It can come out of the blue (or the grey). Open your eyes, and paint the world in brighter colours. Cynicism disappears, and the world seems possible again; seems young again. Love transcends physical weather, bringing its own microclimate of the heart. It can give you spring fever, when it isn't even spring. Make you happy when skies are grey. Love, like sunshine, obeys its own rules, inviting spontaneity and holiday misrule. You are allowed to kiss in public, or don't care that you're not. A lunchtime drink or two on a sunny terrace. Why not? It's not here every day. Neither love nor sunshine.

But if new love is like spring, then can autumn be far behind? Love is a capricious, evanescent entity, as impossible to contain, control and make permanent as the golden stuff itself. The urgent yearning or lamenting for absent or lost love can be very like an eternal winter. 'Ain't No Sunshine When She's Gone' . . . 'The Sun Ain't Gonna Shine Anymore' . . . What better way to express the unutterable loss of love, as the sun leaving your sky and stormy weather taking over?

Weather provides the perfect emotional barometer for love. Every day we are reminded of the capriciousness of a much desired ideal; of something that might start out promising so much, and yet by lunchtime or midsummer the clouds appear. And before you know it, it's raining in your heart. And if it's raining over your head too, then all the better. Do your crying in the rain. Not just to hide your tears, but because you will have nature sharing your despair and painting it across the vast heavenly canvas. Love songs, especially those dealing with its loss, are often drenched in the pathetic fallacy.

You are unlikely to hear Nick Cave's 'Love Letter' (2001) on the radio, which is a shame because it's a staggeringly beautiful song of forlorn love and a broken heart. It is also a pop tour de force in the pathetic fallacy. Cave knows his craft well, and claims that the best love songs 'resonate with the whispers of sorrow and the echoes of grief'. His own oeuvre bears this out, constituting some of the most melancholy reflections on love you are ever likely to hear – 'lifelines thrown into the galaxies by a drowning man', as he describes them.[5] 'Love Letter' tells of one such drowning – in rain.

The lover is on his way to deliver a letter attempting to patch up a broken relationship. It is a prayer for forgiveness. The song is punctuated with references to the storm clouds gathering overhead. Nick Hornby discusses it in his *31 Songs*. He admires the song, mostly for its haunting plaintive melody, but is rather dismissive of its lyrics. For Hornby, 'the conceit . . . is pop-trite, and despite the Hardyesque pathetic fallacies . . . the lyrics scarcely bear the weight of Cave's existential despair'.[6] I disagree. It's precisely because the sentiments are so conventional that the weather has to carry such a momentous emotional burden. The gathering clouds express more eloquently and emphatically what the lover's stumbling words can't. The last lines deliver the prayer that is the text of the letter: simply that she will come back to him. We're not told if the prayer is answered, but we don't need to be.

We can't read the letter, but we can read the skies and what they portend. The weather forecast that has been woven into the song, of swelling, impending torrents, provides a definitive answer that resonates thunderously beyond the song's duration. The clouds burst, and the rain falls, and words, prayers and hopes will surely be washed away.

'Pop-trite'? Isn't most pop trite? Especially love pop. You want pop to lift you higher when you are in love by echoing your bliss; or console you when you are down by understanding your pain. Pop is trite, because love is trite. That hasn't put an end to love, or to the love song. Every generation thinks it is in love for the first time. Caught in that sensual music they are happy to swallow the clichés and the lies which love and its laureates sing.

As spring restores and refreshes the earth after winter, so love restores and refreshes the sentiments and sophistries uttered in its name. Seasonal amnesia every time. Love is begotten, born and dies, and there are songs for every season of love, dressed often in the cloudy cloak or golden raiment of the weather. But this is not a new idea. Such lyrical adornment goes back a long way.

YOUNG MEN'S FANCIES

Before there was pop there was poetry, where many of the rules and properties of love lyrics were established. Sir Philip

Sidney's sonnet sequence, *Astrophel and Stella* (1591), is a storehouse of standard lyrical conventions. Sidney's sequence is written in the character of Astrophel (meaning the star-gazer), who is smitten by Stella, the starry distant beauty who has captured his young heart. Or at least his poetic fancy. We follow his raptures and pains, effusions and laments through 108 sonnets – mostly laments, as love's pain was the principal theme within the conventions of the love lyric. In sonnet 91, Sidney's star-struck lover complains:

> I am from you, light of my life, misled,
> And whiles – fair you, my sun, thus overspread
> With Absence's veil – I live in Sorrow's night . . .

Which is basically 'The Sun Ain't Gonna Shine Anymore', or the Manic Street Preachers' 'You Stole the Sun From My Heart', in flowery Elizabethan verse. A few sonnets later Astrophel laments:

> When Aurora leads out Phoebus' dance
> Mine eyes then only wink: for spite, perchance,
> That worms should have their sun, and I want mine.[7]

An elegantly erudite version of 'Ain't No Sunshine When She's Gone'.

Pop presentiments are everywhere in poetry. Carole King's song 'It Might As Well Rain Until September' is basically a beach-blanket reworking of Shakespeare's sonnet 97 from his own sonnet sequence (1609):

> How like a winter hath my absence been
> From thee, the pleasure of the fleeting year! . . .
> And yet this time remov'd was summer's time . . .
> For summer and his pleasures wait on thee.[8]

If love can be a surrogate sunshine, or its loss eclipse the real thing, then love should be accompanied by its seasonal counterpart. Summer just isn't summer when you're in love and apart. Same sentiments, differently expressed. Does it matter? Of course not. Is it conscious on King's part? Unlikely. It doesn't need to be.

Both love and the weather are hardy perennials of human emotion and experience. You'd think that this would make bringing them together even more clichéd, but somehow it doesn't. There's a measure of stability in the problems of love, as there is in the emotional values attached to the elements, the passing of the seasons, and the vagaries of weather. Weather is reliably unreliable, and so is love. There may be nothing new under or about the sun, but as the sun itself brings eternal renewal the weather is always there to express love's joys and pains.

Shakespeare's sonnets are obsessed with the weather, as befits his status as the national poet. The sonnets that use the weather are among his most famous poems, and therefore some of the best-known lyrics in the English language. They must have entered the lyrical idea bank, and provided useful metaphorical currency throughout the centuries.

They are obsessed with the weather because they are obsessed with time. And time is ever the problem of love. The brevity of life, love and beauty is the main theme of the first sonnets in the sequence. These poems are addressed to, or are about, the mysterious young man whose identity has been the cause of so much speculation. Some of these express sentiments of love according to poetic (and Platonic) convention; but they are mostly concerned with persuading the young man to reproduce. What a tragedy that such beauty and nobility should die with you. So get to it, lad, with my blessing. In his fifth sonnet, Shakespeare employs a highly conventional metaphor to argue his case:

> For never-resting time leads summer on
> To hideous winter, and confounds him there,
> Sap checked with frost, and lusty leaves quite gone,
> Beauty o'er-snowed, and bareness everywhere.

The suggestion that human life, like the year, has its seasons is a well-established and seemingly 'natural' idea. It appears natural largely because it is so very conventional. The metaphors are as dead as the leaves. And yet it provides a succinct way of urging the need to seize the day. And, since we no longer enjoy perpetual spring, it is a lesson that is always relevant, and one which age will always need to preach to youth. For as Shakespeare put it elsewhere: 'life was but a flower . . .'

> And therefore take the present time . . .
> For love is crownèd with the prime,

Life is brief. And youth, like spring, is the briefest, sweetest season, when life and love are at their 'prime'. Life has its 'lusty' flowering, but it soon withers, and is cut down by Time's scythe. That is why love is 'crowned' in spring – with a garland of May flowers – and why spring has traditionally been seen as the time dedicated to love.

Time in Shakespeare's sonnets is nearly always measured by the seasons or the weather. In the second sonnet it is 'forty winters' that will 'besiege' the lover's brow with wrinkles before long. In the sixth it is 'winter's ragged hand' that will deface his 'summer'. Then, of course, there is sonnet 18, perhaps his most famous:

Shall I compare thee to a summer's day?
Thou art more lovely and more temperate.
Rough winds do shake the darling buds of May,
And summer's lease hath all too short a date.
Sometime too hot the eye of heaven shines,
And often is his gold complexion dimmed;
And every fair from fair sometimes declines,
By chance, or nature's changing course, untrimmed . . .

It's a wonder he asks permission to make such a comparison: he'd made it repeatedly already in the sequence. And yet this sonnet is saying something different from his earlier poems concerned with time's ravages. When these poems used pleasant seasons or days to point lessons in brevity, they implied a certain regularity and even dependability in the weather. If you're urging the need to seize the day, that day ought to be worth seizing. Convention relies on certain qualities of a season – fair, sunny, warm – being paired with those of youth – beauty, happiness, lustiness – for this to work. The allegory is exact. If only the weather were.

But it is not, and so Shakespeare uses the fickleness of the weather to make a much more subtle, original, and perhaps valid point. Time is not just about brevity here, but chance, randomness and nature's untrimmed whim. 'Sometimes', used twice, introduces a different note from the regular, orderly motions with which our poet lovers have so far contended. Shakespeare's poem breaks out of the realms of allegory, received wisdom or even cliché, into that of lived experience. The rough winds of reality blow through the May mornings of poetic convention. And so it comes alive in our imaginations, and provides a durable and relevant lesson about love in a cold and unreliable climate. The best-known poem of the English national poet is effectively a metrical moan about the weather.

To compare your lover to an English summer's day or sun isn't that much of a compliment. If you have a complaint, however, or are suffering the pangs of love, then you're in business. For the sun of a typical English summer's day is the perfect analogy for brief, fickle or absent love. Shakespeare's thirty-fourth sonnet demonstrates this perfectly, telling a tale with which we can all no doubt identify:

> Why did thou promise such a beauteous day,
> And make me travel forth without my cloak,
> To let base clouds o'ertake me in my way,
> Hiding thy brav'ry in their rotten smoke?

These lines could so easily be addressed to the weather forecasters who have led us all to do likewise in our time; or to the morning sun itself betraying us with false promise as we select our wardrobe for the day. They are actually addressed to the lover himself, but Shakespeare's metaphor for betrayal or indifference is no doubt gleaned from experience: of the weather at least. To compare a lover's desertion, indifference or simple moodiness to a day spent in soaking clothes, when you anticipated sauntering or lazing about in the sun, is spot on. As valid today as it was four hundred years ago.

So yes, if the weather affects our moods, it can betray moodiness in its own right. Ruskin is right, weather can 'sulk', and never more so than when it is resembling the trials love can put us through. Just how is it such a fair prospect of felicity can so suddenly cloud over, and such rough winds strike up from nowhere? The picnic hamper of love's delights that spilled out so sparklingly not two hours before repacked in sullen, sodden silence? Well, that's my experience, and why these poems and songs speak so eloquently to me.

BOTH SIDES NOW

Looking back on my experience of love over the years, it resembles nothing so much as a long English summer. By which, of course, I mean a short English summer. Typically overcast, with intermittent sunny spells, the occasional scorcher, and even a run of blissful promise now and then. But clouds always got in the way, cutting off hope for days, weeks, years. Alas, my mistresses' eyes have been very like the sun. The English sun, that is.

I will recall two particular loves of sunshine and despair. The first started as winter ended, and blossomed during a beautifully mild early spring. A spring out of a Chaucerian love vision, glowing with the prismatic intensity of an illuminated manuscript. That's how the world appeared to me. To us both, perhaps, but briefly. The sun shone, the flowers came early, and each herald of advancing summer brought beguiling anticipations and plans of how it would be spent together. And then, out of the blue, came the cold snap. Snapping the golden thread of illusion that had bound my hopes so closely to the coming summer.

By the first week in May it was over. On an unseasonably hot bank holiday weekend, she chose to announce the end of the relationship. With London dressed in its sunniest garb, I walked through a park unable to understand how the sun could still be smiling. And smile it did, the whole summer. The weather kept its promise; it was the love that had lied. I spent that summer under a cloud of Nick Cave and gin. So much for the pathetic fallacy.

Many years later I fell into love in late August at the end of an unconvincing summer. The day we met was the very day summer died. I noted the irony with bemusement at the time, but for once my heart was elsewhere. A few weeks after we met, and in the first giddy whirl of courtship, we were separated by a solitary late-summer holiday I'd booked to southern Spain. While I was there, I spent hours indoors, tearing myself away from the Sevillian sunshine to send long, intense e-mails to the object of my affection back in rainy London. And though it rained until and into September, back there, with her was where I wanted to be. A rival sun was glowing and growing prospectively in my heart. My return to London was for once untroubled by the usual pangs of separation from the sun. My descent into the grey at Stansted, and into a winter come early, went unnoticed. What did I care? My new love would keep me warm. A SAD lamp for the soul.

It all turned out right in the end. Not the relationship – oh no, that was doomed. But for poetic convention and the pathetic fallacy. Tempest tossed through a temperamental relationship and another ropey summer, I was coldly, carelessly dropped in chill February. But at least this time the weather understood. Through March, April and even May the sky was as black as my thoughts. If nothing else, this restored my faith in pop songs and poetry, and what they had to say about love and the weather.

And then when it appeared that the sun would truly never shine anymore, the clouds lifted, the sun shone, and on came a summer that defied all auguries, broke nearly all records, and mended my poor heart. It arrived, stayed around, and kept its promises. It understood. For months the sun continued to shine. It dried up my tears, restored my faith, and with it my love. My love of sunshine.

To wake early and see the sunlight peeping through or lighting up the blinds on yet another morning, was like the reassuring presence of a loved one in the bed next to me. As you might press your leg or hand against the sleeping lover's form, so I pressed my heart against its healing, reliable presence, and drifted back into sleep, assured of one certainty at least. That the sun would shine that day.

The summer developed a character (as they always do) just like a relationship. By the end of June a pattern has usually emerged, and you can tell how the summer will pan out. Flighty, uncertain, stop-start and doomed to fizzle out, like the two before; or, like this one, and much rarer, committed and inspiring trust. And so it proved to be. As the grass turned browner, the colour returned to my life.

The weather can therefore be a comfort to those buffeted by love. It delivers daily sermons on brevity, fickleness and precariousness. Or it can lift us up with its surprising, restorative glory, chasing away clouds, drying our tears and bringing wonder back again. As mirror or metaphor to our moods, it helps us understand that such highs and lows are part of the order of things.

With all this in my heart and head, small wonder I thought that if love had a colour it was that of the skies. With forty years of English weather, thirty years of pop songs, twenty years of poetry influencing my views, how could I think otherwise?

SUMMER LOVIN'

Yet this relationship between hearts, heavens and love lyrics is not quite as unified or universal as I've implied. Local differences of time or place shouldn't be overlooked. As Gavin Pretor-Pinney, king of the cloud-spotters, points out: 'the ancient poets of India always saw the start of the monsoon season as a time of great romance. They regarded it as the romantic poets of Europe did the spring.'⁹ In Persian love poetry, a shady garden with plashing fountains and refreshing breezes provides the perfect conditions for hot love. The emotional meteorology of love is nowise universal.

Shakespeare doesn't quite claim it might as well rain until September in his sonnet. In England it so easily could. And although the basic sentiment of King's song is familiar – even 'pop trite' if you will – the imagery and sensibility she evokes are very much of their time and place. Such differences are important, as they help shape the relationship between love and sunshine, sunshine and love, enshrined in pop and taken into our hearts.

George Harrison's 'Here Comes the Sun' (1969) comes from a very different climate to the Doors' 'Waiting For the Sun' from the year before. What – like a few hours, Jim? Sit out in the Californian desert, roll another joint, and it'll be here soon enough. Not the years it can seem like for us. ELO's 'Mr Blue Sky' (1977) is about as British as you can get – 'The Sun Has Got His Hat On' (1937) in beardy-weirdy guise. A mini-opera of grandiloquent euphoria (making Queen look like a bunch of shoe-gazers), it celebrates the simple fact that the sky has cleared, and the sun has returned. It paints a child's picture-book view of the heavens – precisely how many of us view them here.

The Beatles' 'Good Day Sunshine' (1966) neatly brings together the whole theme of this chapter. It's about the joys of being in love, and being at one with the world. And what's more, the sun is shining. It celebrates the successful coordination of two most precarious ideals, and that's clearly worth writing a song about. Lou Reed's 'Perfect Day' (1972) is also a song about a special day spent with someone special. It lists all the little things that make the day so perfect: what they drank, where they went, how it made him feel. But it doesn't even mention the weather. Maybe Mr Reed didn't notice it through his shades or the narcotic haze; but, for my money, not to include the merest mention of the weather is a conspicuous omission from a day declared 'perfect'. Almost unimaginable in British pop.

Seize the (sunny) day appears to have survived as a valid sentiment in British pop, but is perhaps less prevalent in America. Especially if we move from Lou Reed's Greenwich Village art rock back to Carole King's beach-blanket teen idyll 'It Might As Well Rain Until September' from 1962. To declare that you have no interest in the fine weather, and are going to mope all day long indoors writing love letters while your friends are enjoying themselves at the beach, is a fine way to express the pangs of separation from your absent lover. Especially when you can be pretty sure it's not going to rain until September, and you might feasibly join those friends at the beach now and then and do some of your moping in the sun. The main problem for the pining lover here is no longer brevity – as it was in the traditional poetic use of fair-weather imagery – but duration. Having to endure dependable, beach-picnic weather all summer long until the lovers are reunited in September. Truly my heart bleeds for them.

'All summer long'. In Britain those terms sit uneasily together, and could almost be considered mutually exclusive. But in King's song duration proves devotion. Desire is still tangled up with the elements, even when they are at variance. The song is as much a testament to the importance of the beach and the sun in the American Teenutopia, as it is to the eternal torments of love. Such sentiments mark a shift in what could be called the climatology of the love lyric. They are certainly eloquent of a very different sensibility to what we have encountered so far.

Let us situate this sensibility in southern California at the beginning of the 60s. For love, especially young love, was very much in that sultry sunny air at the time. Love and summer went together like rock 'n' roll, Jan and Dean, ice cream and soda, drive-ins and neckings, hot rods and surfboards . . . A poster art production advertising a new pop vision of love and sunshine, this marked a major change in the union of lyricism and love.

As we know, spring was traditionally the season for love in European poetry. This idea has its roots in folk, even pagan, wisdom, and hails from a very different society to the one most of us exist in now – one much closer to the land, where-human life was more aligned with the seasons and more at the mercy of the elements. The link is enshrined in such early English lyrics as 'Spring has come with Love', through Chaucer's celebration of St Valentine's day in *The Parliament of Fowls* (late fourteenth century) up to the May-time festivities depicted by Thomas Hardy in *Tess of the D'Urbervilles* (1891). It is received wisdom that in spring a young man's, woman's, bird's or even bee's fancy lightly turns to thoughts of love.

Though this is the stuff of poetic convention, such sentiments must have reflected a measure of reality. Thomas Hobbes, the seventeenth-century philosopher, claimed that life was 'solitary, poor, nasty, brutish and short'. It was certainly truncated or telescoped by our standards. Youth was a brief hiatus between childhood and adulthood; there was no concept of the teenager. And so the brief period of youth, between childhood innocence and adult responsibility, must have seemed a sweet idyll, and the May Day holiday a perfect time to sport with Amaryllis, or even plain old Daisy in the shade. ''Tis nater, after all,' as Tess's mother sagely observes. If sweet lovers loved the spring it was largely because they had to.

By the late twentieth century this had all changed. The season of love had moved from spring to summer, and time and love started to have a different relationship – at least in pop music. Before the Second World War, and the advent of rock 'n' roll, popular music continued to favour the convention of treating spring as love's proper season. The standard thematic reference guide for pop, *The Green Book of Songs by Subject*, has fewer than two pages devoted to spring and related subjects, with the majority hailing from the first half of the twentieth century.[10] In pre-rock 'n' roll popular music, April is still the time to fall in love in, Paris the place to do it, and the moon the chief luminary to witness it. Classic titles include 'It Might As Well Be Spring', 'Some Other Spring' and 'Spring is Here'.

The British Library's database of sheet music titles of songs published mostly in the UK paints a similar picture. Some of the titles evoke greetings card clichés of sweet melancholy at young love recalled under apple blossom, and suggest that the conventions of poetic love survived long into the increasingly urbanized, technologically enabled century. In those days music was still mostly written by adults, and they seemed content to perpetuate a view of love derived from the formative years of lyrical sentiment. A song from 1936 by Julian Wright tells it straight: 'Spring is the Time to Fall in Love'.

With the triumph of rock 'n' roll and the teenager, this changed. If the teenager wasn't actually born in the post-war period, this is undoubtedly when he came into his full estate.[11] In the 40s and 50s, the teenager became a highly visible, highly marketable, cash-flushed, self-sustaining and articulate phenomenon. With rock 'n' roll culture (including films, magazines and TV shows reflecting and sustaining the new identity), teenagers were given their own voice in pop music.

When pop came to be made expressly by and for the young, it sang about the things that mattered to them. Like cars, and clothes, and drive-ins, and parties, and cash, and hanging out and having fun. The new economy of kicks – teenage kicks. All the things the twenty-year-old Eddie Cochran sang about in 'Summertime Blues' (1958). He's got the blues because he has to work all summer and can't do the things that youth now feels entitled to. He's so indignant he calls his congressman, and even seriously considers taking his case to the United Nations. 'It's just not fair'. Repeat, stamping feet, down the decades that follow.

And, of course, youth sang about love. Mostly about love. No change there. What did change was where, and more importantly when and for how long, love happened. It happened at drive-ins, diners and soda stores, and at the beach. In summer. With rock 'n' roll the summer became a new object of celebration, and expressive of new freedoms, values and time. Teenage extended the duration of youth, that spring-like season between child and adulthood. Both the season of youth and the season of love had a temporal reprieve, an extension to the deadline of the essay entitled 'Responsibility'. Summer was now youth, youth summer, and this had an impact on love lyricism.

Summer had a wholly different meaning to spring. It was

longer, its days were longer, and, if you were fortunate enough to live in somewhere like California where the new teenage summers rolled off the pop production line, these summer days and summer nights appeared to stretch out endlessly. Most importantly, this time was your own. As Jerry Keller sang: 'Well, school's not so bad, but summer's better. Gives me more time to see my girl' ('Here Comes Summer', 1959). There, in a nutshell, is the new ethos, and the new timeframe for love.

Keller's song (which reached number one in the UK in October 1959) is interesting for the way it combines references to the old sentimental world of love – moonlit walks, hearts entwining – with the new itinerary of teenage kicks – drive-ins, meeting the gang at cafés (he even mentions his flat-top hairdo). But it also evokes a new sense of expansive time. 'More time . . . lots more time . . . all the time . . . every day'. Not a brief moment to seize, but one to relax into and enjoy at your leisure. All summer long. An extended term for youth and love. Time was now on young love's side.

A new idea, unheard of in the annals of lyrical love. Love was short, because life was. Time to kill? No, Time killed us. The sunlight on the garden reminds us of this. The shadow moves round the sundial (inscribed *Tempus fugit*), and withdraws all too soon. So we cling to such rare and fleeting moments, and to each other tighter still. To be filling time, or killing time, hanging out and enjoying the long days and hot nights stretching before you and your summer love is a new and seductive idea. From then on, summer belonged to the teenager, love had a new season, and spring was relegated to a warm-up act.

Summer is self-contained and inward-facing. It isn't leading anywhere. It just is. Whilst there is an early summer

All summer long. From the 50s love
had a new season

and late summer, the 'trick' to summer is to suspend time, to
let it hang there in the balmy air and never fall to the ground.
Like a surfer cresting the perfect break, hanging ten for eternity. In that endless, car-roofless summer idyll, clammy with
spilled Coke and desire, you might truly think warm days will
never cease.

The pop archive witnesses this change. *The Green Book*
has four pages devoted to summer and associated subjects
(including dates, sun and school, as in 'School's Out for
Summer'), the majority hailing from the post-war period, and
firmly established by the mid to late 60s. Between 1956 and

66 there were seven songs with the word 'summer' in the title among the top US hits for those years: two more than there had been for the preceding six decades.[12] In September 1966 the US top twenty featured 'Summer in the City', 'Summer Time', 'Sunny', 'Sunny Afternoon' and 'Sunshine Superman'.

But really, the first summer of love occurred nearly ten years before, about 400 miles down the coast. When rock 'n' roll stumbled on to the beach, paddled out into the waves, and created surf music, summer love found its eternal shrine and site of pilgrimage. With the first surf and beach party movies (from 1959), and the Beach Boys turning a local cult into a national obsession (and any hick from Kansas into a potential devotee), summer on the beach became a sacred principle. The *only* thing in life.[13]

As Nik Cohn put it in 1969, writing pop's first retrospective, 'California is teen heaven'. In pop music it

> is the joob-joob land far beyond the sea, where age is suspended at twenty-five and school is outlawed and Coke flows free from public fountains and the perfect cosmic wave unfurls endlessly at Malibu . . . And it has been made like this when kids live in grey cities, tenement blocks and it keeps raining and they know it can't be right, there must be something better. California is the something better.

Verging on 'complete fantasy', it nonetheless nurtured the rebirth of love. If poetic love was born in a Provençal meadow amidst spring flowers, it came of age – teenage – on a summer beach-blanket off Malibu. Teenage Arcadia, the golden age restored and extended. Kissed golden by the summer-long sunshine. As Cohn points out, fun, fun, fun depends on 'sun, sun, sun. Surf in the morning, hotrod later and maybe a barbecue at night – isn't that the way that life should be?'[14] From a season, to an idea, to an obsession – and an endlessly marketable image. Yes, that is the way life should be. The way love should be.

Yet, as I write, the rain pelts down. It has rained all morning, it has rained for the last two weeks, and for most of last month, and it's supposed to be 'summer'. The tenement opposite is looking decidedly damp, and I declare with Cohn that this can't be right. It isn't right. That golden land of love and youth seems far distant, the storm clouds of middle age fast gather, and I think to myself: surely summers were better when I was young . . .

7

MEMORY

What are those blue remembered hills,
    What spires, what farms are those?
That is the land of lost content,
    I see it shining plain,
The happy highways where I went
    And cannot come again.
      A. E. Housman, *A Shropshire Lad* (1896)

The past is indeed a foreign country. They have much better weather there.

My earliest fully-formed memory: I was four, and supposed to be meeting my mother outside the school gates. She was a few minutes late, and I took the opportunity to sneak into the allotment next door. Strictly out of bounds. But the door was ajar, and the temptation strong. I can still recall the ecstasy of tearing through the sun-drenched groves of sky-high scarlet beans and lavender, chasing cabbage whites that tumbled from my grasp into the impossible blue beyond. Most of all, I recall the sensation of being energized by the sun. A photosynthetic surge that intoxicated me then, and ever since.

I don't remember how I got out. Maybe an angel with a flaming sword expelled me and shut fast the door; for, on being retrieved, I apparently declared to my frantic mother that what I had found there was 'paradise'.

I stand by that claim. For I must have felt instinctively the transformative power of sunshine; its alchemical ability to turn the humblest patch of suburban verdure into the Elysian Fields, and how it can preserve and sanctify moments from our pasts. If I didn't know it then, I certainly know it now, and can view this moment as formative in my heliotropic quest. A version of the oldest journey of all – to get back to the Garden.

ET IN ARCADIA EGO

Eden is the Judaeo-Christian version of what classical writers called the Golden Age. A defining myth in Western literature and art, it has equivalents in nearly every religion and culture. Ovid describes this age in Book I of his *Metamorphoses* (first century BC):

> The fertile earth as yet was free, untouched by spade or plough;
> And yet yielded of itself of everything enough . . .
> The springtime lasted all the year, and zephyr with his mild
> And gentle blast did cherish things that grew of own accord,
> The ground, untilled, all kind of fruits did plentifully afford . . .
> Then streams ran milk, then streams ran wine; and golden
>   honey flowed
> From each green tree whereon the rays of fiery Phoebus
>   glowed.[1]

The essentials of this picture are peace, repose, rural simplicity and (best of all) the enjoyment of perpetual spring. It is in truth a hybrid of spring and autumn, combining freshness and innocence with fruitful abundance – the perfect combination, and only achievable in idealized myth. Regret for this blessed time, and an acute awareness of the difference between *now* and *then*, has resounded down the centuries, making nostalgia one of the oldest, most resilient and widespread sentiments in Western literature.

Versions of the Golden Age, or Arcadia (a place that recaptures elements of this age), or Eden (the biblical equivalent of Arcadia, lost through a fall from innocence), are central to the literary tradition known as the pastoral. Though originally referring to a genre of poems depicting the idealized lives and loves of shepherds, the 'pastoral' strain has been adapted over time, and is a strong presence in modern literature – as much in sentiment as in formal literary style. This sentiment is best summed up by the Latin phrase, 'Et in Arcadia Ego' – usually taken to mean, I too have dwelt in Arcadia. It is the implied or explicit epigraph to countless novels and memoirs of the post-Romantic period, and is a conspicuous refrain in English letters from the first part of the twentieth century.

Evelyn Waugh used 'Et in Arcadia Ego' as the title for the first part of his novel *Brideshead Revisited* (1945), a perfect example of the English pastoral sentiment in fiction. *Brideshead* is a very sad book, especially if you don't buy the Catholic affirmation at the end. Being a hopeless heathen and incorrigible romantic, it is the profane rather than the sacred memories which I love most. Especially those from the first chapters where the narrator recalls his days in Arcadia. 'I have been here before,' recalls Captain Charles Ryder, on realizing that his regiment has been temporarily stationed in Brideshead Castle, the scene of many memories.

I had been there before, first with Sebastian more than twenty years ago on a cloudless day in June, when the ditches were creamy with meadowsweet and the air heavy with all the scents of summer; it was a day of peculiar splendour, and though I had been there so often, in so many moods, it was to that first visit that my heart returned on this, my latest.

Ryder is recalling his friendship with Sebastian Flyte, the beautiful but doomed aristocrat who is in love with his own childhood, and who destroys himself in his need to escape the emotional demands of his staunchly Catholic family. But all that is way off. It is the day of his first visit, when they were both at Oxford, that Charles first revisits in memory. Borrowing a car, they drive to Brideshead, and stop on the way for a picnic:

> It was hot enough now to make us seek the shade. On a sheep-cropped knoll under a clump of elms we ate the strawberries and drank the . . . sweet, golden wine [which] seemed to lift us a finger's breadth above the turf and hold us suspended.
>
> 'Just the place to bury a crock of gold,' said Sebastian. 'I should like to bury something precious in every place where I've been happy and then, when I was old and ugly and miserable, I could come back and dig it up and remember.'[2]

It is Charles who digs it up, however: 'remembered with tears by a middle-aged captain of infantry' over twenty years later. The 'peculiar splendour' of the weather is an essential component of this Arcadian vision, and of modern pastoral.

Perpetual spring was a defining aspect of the classical Golden Age. Yet this had a practical rather than an emotional appeal for the agrarian cultures it was originally conceived to delight. It meant the avoidance of the toil and privations attending the seasonal cycle. Sunshine, enjoyed from delicious shade, was not an end in itself. In the Mediterranean world, sunshine was a given, and even avoided in the torrid summer months. But as we in England can never take our weather for granted, we can never overlook its importance in shaping our desires and ideals. And so the ideal of perpetual spring becomes literalized in English pastoral. Pastoral – the nostalgic depiction of an idealized past through a restored connection with nature – is perhaps only possible now through the alchemy of sunlit recollection.

Sunshine has become the essential, perhaps the final, Arcadian ingredient. Maybe you only need sunlight on a garden to gain entry to Arcadia. Infinitely precious, eternally rare, it sanctifies our pasts even as it transforms our fleeting presents. Those moments, real or imagined, few or bountiful, are the pots of gold we all bury.

TEMPS PERDU

Squinting backwards, the past shimmers like a pool of haze on a distant sunlit vista, and is as mirage-like and intangible when subject to objective scrutiny. I was talking to a London cabbie a few years back, and naturally we discussed the weather. It had been a lovely hot autumn day at the end of a pretty mixed summer. Summers were just not like they were when he was growing up, he complained – glorious, hot and long. I told him that actually the official record suggested a general trend upwards, and that since the 90s there had been more exceptionally hot summers than in the 80s, 70s and 60s combined. My comments immediately terminated the conversation. I had challenged a sacred principle.

According to Bill Bryson, the belief that 'British summers used to be longer and sunnier' is one of the 'idiosyncratic notions you come to accept when you live for a long time in Britain'.[3] It was not shared by any of the Americans, Australians or Italians I consulted, but was overwhelmingly subscribed to by the British sample in my research. Whilst many people I questioned couldn't remember what the weather was like two or three summers before (either confusing years or answering 'don't know'), the majority view was that summers were better when they were growing up. Lacking distinct recollection of past weather, they still had a general opinion about it. Weather has an epochal resonance, making a general, vague but significant contribution to the atmosphere of the past. That the weather was better 'then' is like saying things were cheaper then, and people wore different clothes. When you light the costume drama of times past be sure to flood it with mellow golden light.

George Orwell alluded to this trick of memory in his fictional recollections of a vanished England, *Coming Up for Air* (1939): 'Before the war, and especially before the Boer War, it was summer all the year round. I'm quite aware that's a delusion. I'm merely trying to tell you how things came back to me.' Written during the build up to the Second World War, Orwell's novel both explores and reinforces such delusional historical perspectives.

To be fair, Orwell's narrator, George Bowling, strives valiantly not to succumb to sentimentality. He acknowledges that little boys are savages, rejects the Romantic cult of childhood, and concedes that 'if you look back on any special period of time you tend to remember the pleasant bits'. But his final estimation is that 'people then had something that we haven't got now'. And so, 'What's the point of saying one oughtn't to be sentimental about "before the war"? I *am* sentimental about it. So are you if you remember.'[4] As his nostalgia edits out the winters, poor summers and periods of average English weather, so it smoothes away the hardships, inequalities and privations of an historical epoch into a generalized nostalgia for a somehow 'better' order.

And yet, we can't dismiss Orwell's historical perspective too lightly. The Great War did mark a significant historical watershed, and provided a strong impetus for the persistence of pastoral in early twentieth-century English letters. Paul Fussell's brilliant book *The Great War and Modern Memory* (1975) shows how the pastoral became endemic at the time.[5] It helped express the stark contrasts between idealized images of home and the hell experienced and remembered on the Western Front. Fussell quotes Wilfred Owen's poem 'Exposure', describing the gruesome winter of 1917:

> Pale flakes with fingering stealth come feeling for our faces –
> We cringe in holes, back on forgotten dreams, and stare,
> snow-dazed,
> Deep into grassier ditches. So we drowse, sun-dozed,
> Littered with blossoms trickling where the blackbird fusses.
> Is it that we are dying?

Who among us would use the slur 'sentimental' in this instance? Or of Owen's 'Futility', which records the pathetic attempt to revive a wounded soldier by moving him into the sun:

> Gently its touch awoke him once,
> At home, whispering of fields half-sown

> Always it woke him, even in France,
> Until this morning and this snow.
> If anything might rouse him now
> The kind old sun will know.

Will the sun not work its English pastoral charms in the corner of this foreign field? 'O what made fatuous sunbeams toil / To break earth's sleep at all?' But the sun could awaken. As a powerful medium for memory and mythology, it evoked reveries that helped establish the contrast between the stable, eternal past and the uncertain abyss that yawned ahead.

Never such innocence again – the summer of 1914 was a good one, and the fond recall of the kind old sun immediately reinforced pastoral contrasts at the time, and persists in the way the Great War has entered modern memory. That summer has come to sum up and stand for a whole epoch; even eternity itself – the Golden Age just receding from our view.

With a little help from the summer of 1900, immortalized in L. P. Hartley's memorial to lost weather, *The Go-Between* (1953). The heatwave plays a vital role in the events of that summer, or at least in the way they are remembered by Leo fifty-two years later. Through the twin alchemy of recovered memory and rising mercury, East Anglia 1900 becomes a foreign country. Leo, aged twelve, at first hates the heat, but starts to feel an intense affinity with its progress up the thermometer which he checks daily: 'Ninety-four! Perhaps it was a record, a record at any rate for England, where I believed the shade temperature had never reached a hundred. It was my ambition that it should.'[6]

Leo's recollections are based on his diary, and the summer of 1900 is independently verified as a scorcher. Yet, if one swallow does not a summer make, one or two legendary summers should not stand for an historical epoch of summers past. And yet somehow they have come to. The exceptional is more assertive than the unexceptional, or the exceptionally bad. Imagination dwells most on weather won, not weather lost. Especially in English novels depicting a vanished age, like J. L. Carr's *A Month in the Country* (1980).

> Ah, those days . . . for many years afterwards their happiness haunted me. Sometimes . . . I drifted back and nothing has changed. The long end of summer. Day after day of warm weather, voices calling as night came on and lighted windows pricked the darkness and, at day-break, the murmur of corn and the warm smell of fields ripe for harvest. And being young.[7]

August 1920 (which was dry and warm) recalled from a distance of nearly fifty years thus comes to define a personal epoch, a magic moment outside time. Before long we forget that these are meteorological exceptions, and are left with a composite impression that the Edwardian sun took a very long time to set.[8]

But these novels are not just about an age, but about ageing: the personal past viewed from the perspective of maturity. The two are entangled, and the selectiveness of collective memory is perhaps a public expression of a personal perspective. One that existed long before the Great War's fall from innocence. In August 1787 John Byng, later Viscount Torrington, lamented:

> As for the weather, that will never clear up: eternal rain, clouds and chill. Surely summers were different formerly; or is youth the season of sunshine? For then I thought the

summers bright and warm; but now in my age they appear cold and cheerless.

This sentiment, Byng suggests, might be a version of the allegory of the seasons of mankind that plays a key part in the pastoral tradition. If the infancy of mankind was a Golden Age, then each of us was blessed with our own brief glimpse of this vanished time. And if youth is spring or summer, then it must have been bathed in sunshine. This belief appears to have such appeal that it can override reality even as it is experienced. For as Philip Eden points out, 'Byng's miserable August was actually the warmest for four years and the driest for seven.'[9]

The mythical has too strong a hold on our imagination to be challenged by mere facts and statistics. The notion that 'youth is the season of sunshine' would become something of a poetic, and then literal commonplace in the wake of Romanticism. In post-Romantic pastoral, the child replaced the shepherd as the true heir to Arcadia. As William Wordsworth lamented in his famous 'Immortality Ode' (1807):

> There was a time when meadow, grove and stream,
> The earth, and every common sight,
> To me did seem
> Apparelled in celestial light,
> The glory and the freshness of a dream.
> It is not now as it hath been of yore . . .

For Wordsworth, each individual experiences, then loses, a glimpse of paradise. Every life re-enacts the banishment from Arcadia, the passing from Gold to Iron.

> The sunshine is a glorious birth;
> But yet I know, where'er I go,
> That there hath past away a glory from the earth . . .

> The Youth, who daily farther from the east,
>   Must travel, still is Nature's Priest,
>   And by the vision splendid
>   Is on his way attended;
> At length the Man perceives it die away,
> And fade into the light of common day.[10]

Wordsworth's highly influential poem became something of a sacred text for depictions of childhood, and echoes of it resound through all the Arcadian novels I've been considering. For Wordsworth this light, and the vision it affords, was a recollection and intimation of the celestial origins and final destination of the soul. Over time, the sunshine of this 'glorious birth' has been secularized and literalized, but none of its brightness dimmed. All our narrators look about them, and lament a dying of this light. By comparison, the light of common day is decidedly overcast.

A classic illustration of this tendency is found in Kenneth Grahame's enormously popular collection of sketches of childhood, *The Golden Age* (1895). This and its sequel, *Dream Days* (1898), made Grahame's name before *The Wind in the Willows* (1908) secured his lasting fame. As with his more famous work, a small corner of rural England becomes a Pan-haunted outpost of a banished realm. It is the perspective of childhood that transforms it. In a 'Prologue: The Olympians', the adult narrator presents these tales from a world remembered, 'ere the gate shut behind me'. Olympian is his term for the adults who scarcely intrude upon this ideal realm. They are distant because 'they spent the greater part of their time stuffily indoors'. And missed all that sunshine:

> Well! The Olympians are all past and gone. Somehow the sun does not seem to shine so brightly as it used; the

trackless meadows of old time have shrunk and dwindled away to a few poor acres. A saddening doubt, a dull suspicion, creeps over me. *Et in Arcadia ego* – I certainly did once inhabit Arcady. Can it be that I also have become an Olympian?[11]

Very likely. For, despite this magical realm being viewed through the eyes of childhood, it is surely the Olympian disadvantage of years that sees the weather as one of the blessings of this period. Years spent stuffily indoors (Grahame worked at the Bank of England), repeated experience of summers when the sun doesn't shine brightly or sometimes at all, go to make sunlit Arcadianism a product of Olympian regret. Shepherds don't write pastoral poems, and children rarely appreciate the splendour of fine English weather at the time. These are songs of experience, not innocence, and are testimony to the tricks played by memory.

The literature of childhood has its own climate, because it is always written by Olympians. From Grahame to E. Nesbit, Arthur Ransome to Enid Blyton, summer after perfect summer defines the period of youth. In Ransome's *Swallows and Amazons* (1930), the children enjoy cloudless skies for the whole period of their adventure on Wild Cat Island. In the Lake District? I'd sooner believe a boy could fly. The Olympian perspective intrudes when their mother observes: 'The weather can't keep on like this much longer and when it breaks you'll have to come away from the island anyhow. A camp of drowned rats is no fun for anyone. I've tried it. You make the most of your three days [remaining]. We'll come again next year.'[12] Ah, the voice of experience, the tone of regret. Three more days of perfect weather, three more days in Arcadia. So this is not an entire fantasy, they were just jolly lucky.

Next year, it might be different. More like Virginia Woolf's kill-joy classic, *To the Lighthouse* (1927), where rain decidedly stops play for the disappointed child, and for most of the novel. Ransome's tale might be seen as a sentimental child-friendly response to Woolf's grown-up artistry and irony. On Ransome's blessed island, it is fine tomorrow and the next day. Long enough to spin a rattling good yarn, and make us all yearn for the perfect summers of the literary long-ago.

It may not always have been glorious summer when we were young, but maybe it was in the books we read. A series of sunny vignettes strung together on the golden thread of collective memory, compiled of exceptional summers and selective depictions, or just vague, timeless nostalgia. I too have dwelt in a version of that Arcadia, because I'm sure the sun

All our yesterdays. Golden memories on golden sands

shone more for me too. Do we confuse what we have read or experienced? Not quite. But our harvest of literary impressions reinforces and mirrors a similar tendency in our own memories to dwell on the exceptional and edit out the bad – a virtuous circle of collective and personal delusion.

THE VISION SPLENDID

But such tricks can't all be due to poetic convention. If the past, both public and private, comes back to us as a sun-dappled garden, then this may be because there is something decidedly sunshiny about the function of memory. The novelist Vladimir Nabokov describes his own dawn of consciousness in *Speak Memory* (1989):

> Initially, I was unaware that time, so boundless at first blush, was a prison. In probing my childhood (which is the next best to probing one's eternity) I see the awakening of consciousness as a series of spaced flashes, with the intervals between them gradually diminishing until bright blocks of perception are formed, affording memory a slippery hold.

Light provides a metaphor for consciousness, but also helps the author identify his very first 'bright block of perception'. He recalls his first awareness of his own identity and that the people holding his hands were his parents:

> Judging by the strong sunlight that, when I think of that revelation, immediately invades my memory with lobed sun flecks through overlapping patterns of greenery, the occasion may have been my mother's birthday, in late summer, in the country . . .[13]

This is indeed slippery stuff. Are these 'sun flecks' simply those 'spaced flashes' assembled into blocks of perception, like paint marks arranged on an Impressionist's canvas? Is this applied Wordsworth, or can it be taken as a reliable recall of a sunny day one specific summer? A snapshot captured by the impartial lens of memory, faithfully recording climatic conditions? It is impossible to say, but it does show how light can be both agent and object of autobiographical memory. This makes sunlit memories peculiarly distinct, but also highly unreliable as objective records.

The luminosity of memory shouldn't really surprise us. Sight has always been considered the most important sense, as revealed in our metaphors for comprehension. Phrases like 'insight', 'vision', 'seeing what someone means', all point to the ingrained relationship between vision and cognition. Enlightenment was literally that for Plato. His famous parable of the cave from Book VII of *The Republic* sees the philosopher turning his back on illusory shadows and ultimately identifying reality as that which the sun illuminates. Plato's metaphor for the 'upward progress of the mind into the intelligible region' retraces the probable evolutionary development of mind.[14] Sight was one of the earliest functions of the brain to evolve, and maintains strong traces of this development in its myriad functions, including memory.

Jacquetta Hawkes celebrates the extraordinary achievement of eons of evolutionary adaptation, since sunlight first penetrated man's rudimentary receptive cells:

So great is this power in man that he can draw, paint, carve, the remembered scene or thing when it is not before his eyes. What is more (and man himself does not know how), when eyes are closed and he lies in darkness intensely vivid scenes project themselves before his sleeping mind, lit, it seems, by the glow of remembered sunlight.[15]

Hawkes is describing physiological development within a broader evolutionary narrative, so for once we don't need sentimentality to account for this colouring. The mind recollects what it collects, and has a magpie's eye for the shiny stuff.

Hawke's phrase 'remembered sunlight' might be used also to describe photography, a process that makes tangible much of what I've been discussing. Photography means light writing, using sunlight to fix the past. And before Daguerre stamped his ego on the process by calling his creations 'daguerreotypes', the rival claimants to this invention were happy to give its true progenitor its due. Joseph Nicéphore Niepce, who was Daguerre's partner before his death in 1833, was creating ghostly 'heliographs' from the mid-1820s, and Henry Fox Talbot called some of his own experiments 'sun pictures'. Before the invention of flash, photography's ability to capture a moment in time was wholly dependent on the sun.

Talbot was a keen advocate for the democratic immediacy of this new art form. His researches, from the early 1830s, were born out of his shortcomings as an amateur draughtsman of the picturesque. He wanted to capture a scene while on his honeymoon in Italy, but his talent wasn't equal to the view. So he sought for, and perfected (albeit just too late to claim the inventor's laurels), a means of fixing the scene before him without the need for artistic instruction, or, as he claimed, talent. For Talbot, with the right kit and the right conditions, everyone could be an artist. The true artist was the sun. His *The Pencil of Nature*, published from 1844, carried the following notice in its later instalments:

The plates of the present work are impressed by the agency of Light alone, without any aid whatever from the artist's pencil. They are sun-pictures themselves, and not, as some persons have imagined, engravings in imitation.[16]

The process appeared so magical and unprecedented that sceptics refused to believe that 'nature' alone could achieve such accuracy and artistry.

Talbot emphasized the passivity, even the randomness of the camera's gaze. It was, like the sun itself, impartial: 'It chronicles whatever it sees, and certainly would delineate a chimney-pot or a chimney-sweeper with the same impartiality as it would the Apollo of Belvedere.' The sun god Apollo had been tamed and levelled by his own agency. Talbot was a highly respected classicist as well as an entrepreneurial scientist, and so had an acute sense of historical time and its ravages. In his first published account of his efforts (rushed out in January 1839 to achieve some share of the limelight in which Daguerre was basking), he discussed 'The Art of Fixing a Shadow'. Photography, he admits, partakes 'of the character of the marvellous', for

The most transitory of things, a shadow, the proverbial emblem of all that is fleeting and momentary, may be fettered by the spells of our 'natural magic', and may be fixed for ever in the position which it seemed destined for a single instant to occupy.[17]

The sun, for once, serves eternity rather than mutability. The moment is finally captured in sunlight's nets of gold (or silver iodine). In the magic amber of Talbot's sun pictures we discern a fragmentary gleam from the Golden Age.

And yet, for all Talbot's democratic protestations, photography's dependence on the sun means it is far from impartial in one important sense. Like so much I have encountered in my journey, it is subject to the vagaries of climate. In 1843 Talbot took a tour of northern France to record architectural gems for his forthcoming publication. But he was frustrated by 'weeping skies' (the pathetic fallacy again), and came back with only a few examples. In September that year he visited Oxford, from where he writes to his wife, 'The weather has been exceedingly fine both Monday, Tuesday and today, and I have made about twenty views each day, some of which are very pretty . . .'[18] The early archive of this revolutionary art form is therefore a record of lost weather. A day of splendour, a break in the clouds preserves the long-vanished past for our eyes. The past steps out of the shadows through a trick of sunlight: for the very first time.

We take such tricks for granted now, but we shouldn't. I can't help sharing Talbot's and other early enthusiasts' wonder at how sunshine captures the past, and seeing 'The Doings of a Sunbeam' as a peculiarly apt metaphor for human memory. The phrase is Oliver Wendell Holmes's, and was used by the psychologist and amateur photographer as the title for an article on 'this business of sun-picturing' from 1863. By now photography was a popular hobby, and Holmes celebrates its deeply personal value for the amateur practitioner. As well as immortalizing loved ones, it might capture the image of

> an old homestead, fragrant with all the roses of his dead summers, caught in one of Nature's loving moments, with the sunshine gilding it like the light of his own memory.[19]

Photography is here both an expression of memory and an aid to its persistence and sentimental potency. The gilding of sunlight is both the agent and subject (or perhaps the 'atmosphere') of such recollections; 'sun-picturing' makes the analogy peculiarly apt and exact. From its invention to the latest format for storage, digital 'memory cards', photography provides one of the most compelling metaphors of memory.[20]

Despite Talbot's suggestion that the sun eclipsed the artist as the faithful delineator of reality, he still acknowledged scope for creativity in his new art form. His annotations to a sun picture called *The Open Door* even suggest a new aesthetic. Obviously pleased with the effect created by a shaft of bright sunlight partly illuminating a dark interior, he refers back to the Dutch Interior school of painting:

> A painter's eye will often be arrested where ordinary people see nothing remarkable. A casual gleam of sunshine, or a shadow thrown across a path . . . may awake a train of thoughts and feelings, and picturesque imaginings.

There is still a role for the artist's vision and discernment, even with this most democratic art form. It can transform the momentary into the monumental (with the right eye guiding the mechanical one). Such an aesthetic allowed photography to emerge as an art form in its own right, but it also laid the foundations of much creative expression thereafter. Impressionism, obsessed with recording the momentary minutiae of atmospheric effect, realized Talbot's speculations: capturing precisely the casual gleam of sunshine, the shadow thrown across a path. The movement can in part be seen as a reassertion of the artist's autonomy in response to photography's solar challenge. If Talbot hadn't already named his method 'sun-pictures', then the Impressionists might legitimately have used this phrase themselves.

But not them alone. With Modernism, the ephemeral becomes the epiphanic, and a stray shaft of light can be the trigger for a train of picturesque imaginings. Especially when associated with the past, and used to restore that moment for art. Talbot's comments recall an important passage from that most monumental of memorialists, Marcel Proust. It is found in Proust's preface to his collection of essays, *Against Sainte-Beuve*, and might be taken as the prototype for his theory on involuntary memory developed and dramatized in *A la recherche du temps perdu*. Lost time was also remembered weather, it would appear.

> Passing through a pantry the other day, a piece of green cloth stopping up a broken pane of glass brought me up short, inwardly listening. *A summer radiance* came to me. Why? I tried to remember. I could see wasps in a shaft of sunlight, a smell of cherries on the table, I could not remember . . . I hesitated amongst all the indistinct, known or forgotten impressions of my life; it lasted only for a moment, soon I could no longer see anything, my memory had gone back to sleep for good. [My emphasis][21]

Like Nabokov's 'flashes' of memory, like Talbot's stray gleams or Holmes's 'gildings' of sunlight, the past appears to be both preserved in and accessed through light. The mind's eye catches the merest glint of a recollection, and pursues it to its source through the caves of oblivion; or clutches at the motes of memory, swirling in a stray shaft of illumination but refusing to take distinct form. 'Summer radiance' suggests a vague aura of pastness, intangible but charged with emotion – the mere ectoplasm of happiness past. Sunlight appears to have some special role in the process of involuntary memory which Proust explored so exhaustively.

FOOL'S GOLD

Why is this? Is sunshine itself Proustian, able to trigger involuntary memories? It is for me. It only takes sunshine on gera-
niums to recall the hot dusty aura of the glass porches of
seaside guest houses of thirty years ago. Or the smell of hot
rush matting garnished with sun cream to evoke countless
foreign beach jaunts subsequently. For warm fig trees to take
me back to Menton 1986; jasmine, to Córdoba 2004. The sun
has a potent charm in restoring such moments. There is a red-
brick Edwardian school converted into flats next to where I
live, just like the one I attended to the age of eleven. In the
intense heat of summer 2003, I was walking past it and the
smell of the bricks baking whisked me back to the playground
of my childhood. Playtime regained, through this sensual
seance of summers long past.

I recently had my first (and I hope last) experience of
sleeping in a tent since I was very young. I was attending my
first ever music festival, aged thirty-seven. I don't like mud. I
do like warm showers and comfortable beds. End of story. But
I was lucky. A glorious weekend out of a ropey year, and my
borrowed tent was more than adequate. Sunlight flooding the
tent awoke me to the summer mornings of my childhood
holidays. To the shadowplay of grass pressed against the sides
of the tent, of elongated figures picking their paths over guy
ropes on their way to the washrooms, to the mumble of
drowsy campers, and smell of bacon frying over Calor gas. All
restored to me by the play of light against canvas (or whatever
tents are made of now).

Sunbathing in southern Spain a few years ago, I caught a
whiff of a certain suntan lotion, or maybe it was a bodily

memory itself – as Proust himself put it, 'our legs and our arms are full of torpid memories'[22] – and I suddenly recalled my very first experience of deliberate sunbathing. Thinking back, I was pleased to pin it to the glorious summer of 1976. I was nine, and on summer camp with my school. Worn out by the morning's savage boisterousness and succumbing to the heat, a group of us flopped down in the grass. A particularly stern teacher proceeded to apply suntan lotion on herself, and I thought I'd join in this adult caper myself. Requesting some, I was told there was no point putting cream on until I had 'exposed my puny body to the sun'. It's a wonder I wasn't traumatized by what would nowadays contravene best medical advice. But I did expose, and I did apply, and no doubt luxuriated in my new-found hobby as the sun beat down from a peerless blue sky in that peerless summer. Its memory unlocked by a stray sensation thirty years later.

Yet, despite these Proustian transportations, such moments are the makings of shared delusions. For, as it turns out, I was wrong about the year of my initiation into the rites of sun-worship. That particular sense key unlocked the wrong door. Verifying the date with the school records, I found it was not the glorious summer of 1976 when I went on this summer camp, but the following June. I checked the Met Office records for that month, and learnt that the sunshine for June 1977 for the area was decidedly below par. Whilst June 1976 enjoyed 266.8 hours of sunshine, and an average temperature of 25.4°C, the following year only mustered a meagre 126.7 hours of sunshine, and a feeble 17.2 degrees. The national average for June is 168.4 hours of sunshine. What I had remembered as a representative day from a legendary summer was, in fact, an isolated sunny spell out of a dreary month. No wonder the adults were making the most of it, and dedicated that afternoon to sunbathing.

More typical than legendary, I had mythologized my own past. And yet, according to Trevor A. Harley, such tricks are themselves highly typical. An academic psychologist, Harley is also a weather obsessive, who runs a website about the British weather. He has published a fascinating essay on the role of weather in memory, pointing out that people's 'nostalgia for a particular weather event is [often] based upon statistically incorrect data'. As he reasons:

> It is likely that the source of this incorrect belief is a combination of unquestioning subscription to a popularly held myth, and generalization from one or two prominent but unrepresentative examples.[23]

It's a fair cop. I'd transformed an exceptional day in a ropey month into a typical day of an exceptional year, and devised my own personal myth on the back of it. Harley explains how memory can make 'prototypes' out of scattered, and not always representative, weather events, creating an emotional average that doesn't bear objective scrutiny. It works with white Christmases as well as superior summers, and before you know it, a rare blanket of snowfall, or an exceptional season, month or even day, has led us to believe winters were whiter and the summers more golden when we were young. The statistical spikes leave the keenest impression, pierce more deeply the fabric of memory.

It is the emotional or sensual intensity of these moments that ensure their selective preservation. Intense brightness, intense warmth are likely to leave a more vivid impression than the general slew of grey, however representative. The first day of summer holidays, the sun is shining, you run around the garden hugging yourself with glee that there are six whole weeks of freedom stretching ahead of you. No matter if that very afternoon the clouds rolled in and the sun hid his face, scarcely showing himself again for the rest of the holiday. The emotional peak of that first moment has etched its trace. Sunny memories glow with vivid intensity, and expand to fulfil an emotional need. But were we happy because it was sunny, or do we remember it as sunny because we were happy? Does the mind remaster the tapes of memory, editing and recolouring them according to our emotional demands? It is surely significant that Charlie Kaufman and Michael Gondry's 2004 film fantasy about erasing unhappiness from individual memory is called *Eternal Sunshine of the Spotless Mind*. Taking the title from Pope's couplet, sunshine is used as a metaphor for blissful untroubled memory. The spots of pain erased out, the clouds airbrushed from memory. Testimony to sunshine's potent role in selective sentimentality.

But perhaps there is a more mundane explanation for the collective delusion of childhood summers – the simple fact of greater solar exposure. I don't know how it is now, or what it was like for you, but my school hours were 9.00 a.m.–3.30 p.m. There was morning play, lunch hour, and afternoon play to boot (and if it rained they kept us in). Then there was a six-week summer holiday. As adults, we spend the majority of our working lives indoors (about 90 per cent of our time, according to Richard Hobday, the expert on sunlight and health). And if you're like most people I know, and work far too hard, you're lucky if you see the sun during the week. That leaves weekends, which by some evil twist of fate are often the worst days of the week for weather. So let the meteorologists produce their sunshine statistics, and tell us we've never had it so good. That's cold comfort if those hours were all totalled while we were stuck indoors. It's weekend, bank holiday and holiday sun that counts. A good or bad summer is an experiential rather than a mathematical assessment, and I refuse to be persuaded otherwise. The sun may not have shone more when we were young, but we certainly got more exposure to it when it did.

A friend of mine is adamant that the weather was better when she was young. For she clearly remembers the special birthday teas she enjoyed in her sunny garden in early May each year. Whilst it was always sunny then, it's often touch and go now; and, as she takes the day off work each year she often has to console herself for a dreary day with some retail therapy. She mentioned this to her mother, who revealed the reason for this startling reliability. The birthday tea was a moveable feast, depending on fine weather. If the day itself was poor, then the sunny celebration would be put off until the next fine day. Even when our mind isn't playing tricks on us, our parents are.

Photographs themselves might also encourage a sunny view of our pasts. Raiding an old biscuit tin of childhood photographs, I find the majority are taken in reasonably bright conditions. Does this prove that the sun shone more then? It's more likely due to the fact that sunlight lends itself to a more pleasing photograph (especially in the days before digital cameras when people were more selective in their snapping). Most of these shots were taken on holiday, and no one wants photos of damp and grumpy kids stuck in the soggy tent or the teashop. But I know for a fact there were such occasions. I can remember sitting in a parked car in a campsite, as the rain came down in buckets. I was reading an Enid Blyton book about exciting seaside capers for bright-eyed, strong-jawed chaps in perfect summers, and feeling a profound sense of injustice. I voiced this resentment to my brother, wishing such exciting things would happen to us. But how can they when you're stuck in a flooded field somewhere in Dorset or Wales?

The author in 1969. Cheer up,
the sun'll come out tomorrow

Obsessed with sun from an early age, I'm cursed with a memory that refuses to subscribe to the general delusion. The gap between the ideal and reality, and its attendant disappointments were keenly felt at the time, and rankle still in recollection. I recall trudging round a department store in Bournemouth, in the summer of 1974. How do I know this? Because while we were buying waterproofs to protect us from the deluge descending on the 'English Riviera', a pop song called 'Beach Baby' by First Class (who?) was playing over the radio. A British tribute to the Beach Boys' Californian teen idyll, it was a piece of pure bubblegum. Worse, second-hand bubblegum, evoking the endless summer clichés of ten years before. But it didn't fool me. There we were. The beach outside was a midden of grey soggy sand, our needs were sou'westers rather than sun cream, and I was going to have a right old strop at the disparity. Exactly where, I wanted to know, was this sun he was singing about?

BEXHILL REVISITED

And yet the Great British Seaside resort institutionalizes the association between childhood and sunshine. It is part of its official, that is to say its heritage, identity.

Bill Bryson recalls the effect British children's books had on his expectations of the British seaside. 'They all had titles like *Out in the Sun* and *Sunny Days at the Seaside*', and 'strangely influenced' him, encouraging him to take his

> family holidays at the British seaside on the assumption
> that one day we would find this magic place where summer
> days were forever sunny, the water as warm as a sitz-bath,
> and commercial blight unknown.[24]

Did he find it? Of course not. It doesn't snow at Christmas either, but it doesn't stop us wishing it would, or advertisers pretending it does. A version of this summer mirage shimmers on the horizon of all our memories (if we are of a certain age), and still captivates us. We can realize something of it, if we're lucky or choose the right day. Timing is all, on this small Island.

Seaside and scorcher go together. So when the Bank Holiday sun shows its face, we flock to the coast like migrating birds obeying some instinctive call. Some atavism of cultural conditioning and collective nostalgia asserts itself, and tells us to be beside the sea on a day like this. And if we only visit on such days, and avoid the solar lottery by attempting a full fortnight stretch, then the unique anachronisms of the British Seaside (Bryson's idyll) can be enjoyed in their true element. So take a bucket and spade – the shops still sell them here – and dig up the past; buried treasure from your own and generations of summers past, retrieved and restored with every ritualistic revisiting.

Holiday institutionalizes this recovery, and the beach is sacred to this ritual of restoration. From Georgian, Victorian, Edwardian times to our own, sand called for the discarding of shoes and socks, if nothing more. A space sanctioned to sensual discovery and freedom. For toes of children young and old to be dug deeply in. Recover childhood, recover Arcadia, preserved beside golden, sunlit shores.

Yet a brief glance at the archaeology of beachside ritual reminds us that, for all its faint echoes of eternity, the sunlit component of this idyll is of relatively recent vintage. The waves broke, and the sunbeams fell on these sands for centuries before we children came to lap them up along with our ice lollies. Before then, there were other diversions if the sun forgot to shine as brightly as it does on the posters and in the picture books. It's significant that 'Oh, I Do Like to be Beside the Seaside' (1907) doesn't even mention the sun. Of all the attractions, it was the least important.

But that all changed. It is ironic that a visit to a seaside resort like Bexhill, with its startlingly Modernist pavilion constructed in 1930 as a temple to the new cult of hygienic sun-worship, evokes nostalgia rather than its original spirit of futurism. Sunshine, the agent that brought about a revolu-

Such solar propaganda got us hooked – a warm up act for package tours abroad

tion, now preserves it in memory. From curative to curator – in a living museum of summers past.

In the posters and advertisements for seaside resorts from the 1930s onwards, we can find the seeds of the discontent that would soon send us in search of sun loungers further afield.[25] Few of the resorts advertising in *Summer Holidays in the British Isles*, issued by Thomas Cook in 1950 (the year the mass package holiday abroad was born), neglect to imply some pre-eminence in solar performance, conspiring in the collective imposture that duped Bryson. We are invited to 'Prestatyn Holiday Camp – For the finest holiday under the sun!'; or 'Morecambe and Heysham On the Sunny Lancashire Coast'; or to Felixstowe 'for sunshine on the sunny Suffolk coast'; or to 'Come south to sunny Southsea'. There's 'Something for everyone in sunny Southport'; or we might even try 'Ulster . . . for those who . . . like to sunbathe on fine sandy beaches . . .' Some resorts went even further in their claims. Worthing dubs itself 'Britain's record sunshine town'; Eastbourne the 'sun-trap of the south'; St Mawes claims to be 'the warmest and most sheltered spot in England'; and Weymouth to enjoy '300 hours per annum more' sunshine than London. Such propaganda promoted what they could not guarantee. With predictable results.[26] Sunshine periodically restores what it ultimately destroyed – the bygone era of the domestic seaside holiday.

FOLLOW THE YELLOW BRICK ROAD

In summer 1974 I was hearing the seductive strains of 'Beach Baby' as we dived for waterproofs in rain-drenched Bournemouth; summer 1984 I was in the South of France with my mates on my first independent trip abroad. I have never spent more than a day, and always a sunny day, at the English coast since. In this I conform perfectly with demographic trends. It was in the early 80s that foreign destinations overtook domestic for the main holiday for the first time in Britain. But before there were the means there was the desire. Myths about what summers were like in the past merge with and are sustained by beliefs about what they should be like now, in the future and always. False memory is reinforced by false expectations. But where on earth did these ideas come from?

In February 1963, when Britain was suffering the coldest winter of the century, Cliff Richard's film *Summer Holiday* was showing to snowbound cinemas. The film opens in black and white at an English seaside resort in summer. Old people doze in deckchairs on nearly empty beaches. The skies are as grey as the film stock. A Salvation Army band plays to a few bemused stragglers. But then the thunder rolls, the rain comes down in torrents, and everyone runs for cover. The titles continue with a succession of typical summer scenes, including rained off cricket and browned off young people. Summer cancelled for yet another year.

We do still like to be beside the seaside, but lately we've got a bit more demanding about what this should entail. The titles end with Cliff's chums staring out at the rain from the bus depot where they work. Doomed to damp despair in Blighty, they anticipate the inevitable washout of their annual holiday. But hope arrives in the form of Cliff in a double-decker bus they will take 'abroad' for the first time. Immediately the film turns to colour, as they turn their back on England and its terminal Victorian grey. The future of summer is in Technicolor, the future of summer is abroad.

It is no accident that the great myth-maker of the modern age is situated in and depends upon a warm sunny place, southern California. Cinematic sunshine lights the stage set of our dreams, desires and aspirations, flooding them with chromatic richness and clarity and painting the brick road to happiness a bright sunny yellow. We all daydream in glorious Technicolor, and a Cinemascope sunset is what we hope to ride into one day. Small wonder we yearned for better weather, and can perhaps attribute part of our new-found disgruntlement and disloyalty to the seductive visions of Hollywood. Like Cliff, we went to see if it really was true. It was; the sun does shine brighter beyond these isles; the sea really is blue. California dreaming. First stop Benidorm.

Both 'Beach Baby' and *Summer Holiday* were hymning the same ideal. The latest and final Arcadia. The eternal beach – mythologized in surf and beach movies, Beach Boys' songs and a thousand pop cultural reproductions thereafter – offers a glimpse of the Golden Age restored. Why have eternal spring when you can have Endless Summer? *The Endless Summer* was the title of one of the first authentic surf movies to reach a broader audience (1966). As its poster declared: 'On any day of the year it's summer somewhere in the world'. The film follows

Why have eternal spring when you can
have endless summer? The beach as the
final Arcadia

two young Californian surfers who pursue the sun around the globe, and invites us to 'Share their experiences as they search the world for that perfect wave which may be forming just over the next horizon'. With wide-eyed innocence they stumble across virgin beaches that have never been surfed. For, as their counterparts discovered in Vietnam, 'Charlie don't surf'.

Apocalypse Now (1979) might be seen as the dark twin of *The Endless Summer*. Coppola's film is the consummate statement of the importance of beach culture in the American teenage imagination. They are bronzed beneath their uniforms, and the beach is there beneath the killing fields. So they surf even as they kill, and sunbathe on top of the boat penetrating the very heart of darkness.[27] In the Great War Wilfred Owen had wondered about fatuous sunbeams toiling among the frozen hell of the Western Front. Pointless if they cannot restore a life snuffed out by barbarism. Fifty years later those rays are not for life but pleasure, and they are certainly not going to waste.

With its vast wildernesses and youthful optimism, America has a different take on the pastoral from the English tradition. A small, shrinking country, its empire and glories in retreat, the English tradition seeks stability through preservation. Shoring up fragments of Arcadia against time's ruins. The beach as Arcadia speaks of discovery rather than desperate preservation. The restlessness that characterizes so much of American literature and cinema leaves its footprint on its own, the latest, version of the pastoral. From *Huckleberry Finn* to *Easy Rider*, youth discovered and expressed itself by pursuing the ever-retreating horizon. In the post-war Teenutopia, the beach came to symbolize the ultimate realization and destination of this youthful questing. The beach is the final Arcadia, where only the young and beautiful are really welcome.

Midnight Cowboy (1969) sums up this quest: to go where the weather suits poor Ratso's Cuban shirt. In his daydreams of Florida, the starving, crippled Brooklynite imagines himself made whole and happy by the all-year restorative sun. The poster of Florida oranges on his rat-infested New York squat captures the dream of the new Golden Age: eternal summer and effortless fruitfulness realized by golden sands. The eternal beach has entered modern myth, and is the final Arcadia: a paradise not entirely lost. With mobility and prosperity, cheap flights and a shrinking globe, the dream of endless summer can belong to us all.

Oxford Street, London, late June 2007. The rain is falling in sheets. The longest day was yesterday. I marked the occasion by staying up all night with a band of bedraggled druids to witness the bank of cloud behind the heelstone at Stonehenge become slightly less grey. Truly cosmic. It has rained every day for about the last two weeks. Glastonbury is diluvian, a black cloud is about to deposit its payload on Wimbledon, and Yorkshire is under water. They say more is on its way.

In the fashion shop windows, however, it is high summer. Fair youths in urban beachwear flirt with maidens in bikinis. They lounge in sylvan bliss, about to enjoy a picnic. Real people rush past them swaddled in wet winter wool. As I stare at this parallel universe, a bus shoots past too near to the lake that was once the gutter, sending the contents halfway up my legs. On its side an advertisement for cider shows sunlit summer orchards and tells me it's 'Time to cool down'. My expletive is clouded in my vapoured breath.

In advertising, marketing and in-store promotion they have proper, dependable seasons. Beach-blanket bonanzas in the summer; reliable snowfalls at Christmas. Every December the adverts, trailers and window displays perpetuate a Dickensian delusion of Christmas, last experienced about twenty-five years ago. This pure fiction (the Dickensian Christmas itself was a myth even when Dickens popularized it, based on his own memories of anomalous snowfalls in his youth)[28] has its summer counterpart. According to Philip Eden, records show it is more likely to be a white Easter than Christmas; whilst 'flaming June' is yet another flaming delusion. 'Looking back through the twentieth century, only ten Junes could reasonably be described as predominantly warm and sunny, which leaves ninety which weren't.'[29]

The delusion of summers past sustains the dream of the perfect ones yet to come. As John Berger puts it, 'Publicity is, in essence, nostalgic. It has to sell the past to the future.' We accept these images, Berger claims, as 'we accept an element of climate' – or expect the airbrushed, idealized version of climate.[30] We are sold the dream of perfect summers as we are sold the dream of eternal youth. A complete lifestyle package. The sun always shines on TV. On cinema, in-

Reality. A Brighton bank holiday in the 1940s

store promotion, direct marketing, billboard, you name it. Wherever people are glamorous, happy and fulfilled, the sun smiles blissfully down on them. Spring is accompanied by 'fever', summer deals 'sizzle', and we are kept at a pitch of anticipation for those lazy, hazy, crazy days of summer. The only hazy thing is our memory, and the marketeer's grasp on reality.

OK. I'll take it. I have the money, I have the leisure time. I want the product. So where is it?

8

LIFE

Is it so small a thing
To have enjoy'd the sun . . .
Matthew Arnold, *Empedocles on Etna* (1852)

UNDER HATCHES

Where's summer? Not here, not this year, that's for damn sure. Early July and it is still raining. Summer 2007 is a mere technicality of date. The highest temperature for the last two days has been about 15 degrees. There have hardly been three consecutive days of sunshine since April, when we had a freakish spell of warm bright weather for most of the month. That is now but a distant memory. The barbecue rusting in the yard, the electric fan by my bed, start to resemble archaeological artefacts whose intended use is increasingly difficult to fathom. And so I mope, chief mourner for a stillborn summer. The very heart ripped out of my year.

So I lost a summer, not my life or my property. What does one lousy summer add up to in the scheme of things? We had a good one last year, and a lovely spring this year. It might pick up soon (it didn't). Well, we might have another scorcher next year. Bad though it is, it is hardly unprecedented. The year 1816 has entered legend as 'the year without a summer', or 'eighteen hundred-and-freeze-to-death'. The weather had devastating effects around the globe that year. Failed harvests caused food shortages and even led to riots. The next two summers were not much better, it would appear. By April 1818, John Keats was writing to his friend J. H. Reynolds, 'It is impossible to live in a country which is continually under hatches – Who would live in the region of Mists, Game Laws indemnity Bills &c when there is such a place as Italy?. . . Rain! Rain! Rain!'[1]

When Keats claimed it was impossible to live under these conditions he was perhaps speaking literally. He was nursing his brother Tom through tuberculosis, and they had come to Devon for better air. But it would appear to have rained every day they were there: 'I wanted to send you a few songs written in your favourite Devon – it cannot be – Rain! Rain! Rain!' A few weeks later he was still complaining: 'We are here still enveloped in clouds – I lay awake last night – listening to the Rain with a sense of being drown'd and rotted like a grain of Wheat . . .' The good weather and health they sought persistently eluded them: 'The Climate here weighs us [down] completely – Tom is quite low spirited . . .' His brother died later that year, and not long afterwards Keats's own health went into sharp decline.

When Keats eventually made it to Italy in the autumn of 1820 it was far too late to save him. He died the following February of tuberculosis, aged twenty-five. As Jonathan Bate observes: 'Keats was hurried to his death less by the reviewers [who savaged his poems] as Byron supposed, than by the weather.'[2] Three lost summers of that short brilliant life spent under hatches: that is truly tragic.

Compared with Keats's day, we live very sheltered lives. We have warm, waterproof clothes, and travel in weatherproof boxes; we heat our homes, and spend much of our lives sequestered from the world outside. Out of necessity in our workplaces, and out of choice and habit when the elements drive us inside. If an Englishman's home is his castle, the weather is a major factor in maintaining its fortifications. Our insularity is partly insulation. British TV is widely heralded as the best in the world, providing compensation for, and instruction in, a life lived behind a raised drawbridge. There are programmes about property transformation, gardening, cooking, or about buying second Castles in Spain or wherever the sun shines. In Keats's day it may have been impossible to live permanently under hatches, but since then we've contrived ways to make the great indoors a bit more bearable.

What makes it bearable in the final analysis is our ability to escape it. Keats's trip to Rome was a last desperate bid for life, involving significant outlay, a long crossing, and ten days in quarantine at Naples cooped up in a stifling boat. Now planes take off every minute, from all parts of the country, to whisk us beyond the weather, sometimes for less than the train or cab fare taking us to the airport. The British Isles is the undisputed leader and hub of the budget airline industry. In 2005 the market research company Mintel estimated that the low-cost airline sector carried 80 million passengers around Europe the previous year, of which 60 million started or ended their journey in the UK. Easyjet confirmed to me that the majority of their seats are bought by British nationals. There aren't many things the UK does cheaper and better than our European counterparts. But escaping must be a fair candidate.

It is precisely because we live so much indoors that we set so much store by those moments when we can break free. We may live indoors, but we most fully *live* outside. Society leaves its palaces for the Chelsea Flower Show, Ascot, Henley and Wimbledon. There are the village fêtes, garden parties and summer weddings of the middle classes. And there are the barbecues, picnics and beer garden bacchanalias of just about everyone. Every summer new music festivals spring up to swell an already bursting season of outdoor release, and offering further provocation to the spiteful weather gods. When we do have a decent summer, like 03 or 06, the weather opens a door through which we catch a glimpse of a different life. Summer as continuous narrative, not measured out in egg spoons, or simply cancelled like this one.

It's all very well considering the scheme of things, but which of us is in the scheme of things? We're in our own schemes, and the weather so often stops them happening. Life is what happens while we're waiting for the weather to improve.

And improve it might. As I write, the reality of global warming is finally hitting home. Scientists are being listened to, and recent events have compelled even the most sceptical to accept a link between what we pump into the atmosphere and what the atmosphere flings back at us. Whether political action will match rhetoric and prevent global catastrophe, only time will tell. It goes without saying that climate change is serious stuff, and a highly complex issue. It presents a minefield for the half-informed, forced to rely on media misrepresentation, and who have difficulty thinking both globally and long term. That's me. But it's a question that has an obvious impact on our relationship with sunshine in the future. I can only judge from what I've read, from what I've experienced, and what I feel.

According to what I've read the UK, like the rest of the world, is getting warmer. The greatest increase in mean temperatures and the incidence of heatwaves has occurred during my lifetime. A Met Office press release issued in May 2007 puts it succinctly:

> Since the mid 1980s the average UK surface air tempera-ture has warmed by about 1°C, which is about twice the global warming trend averaged over all land areas. As a con-sequence, summers (and indeed other seasons) warmer than the 1971–2000 average are now common (the last summer with temperatures below the 1971–2000 average was nearly 10 years ago in 1998). Moreover, the underlying chance of exceptionally warm summers, such as experi-enced in 2003 or 2006, has increased.

My experience confirms this pattern: the 80s I remember as hit and miss for summers (like its music), but the 90s as more consistently hot and sunny. There was an exceptional run of summers from 94 to 97, and then intermittent heatwaves, stinkers and a few must-try-harders thereafter. The year 2006 saw a lousy spring in the UK but then, with an exceptional summer, turned into the warmest on record. In 2007 the pattern was reversed: the warmest April on record, followed by the wettest May to July in England and Wales since records began in 1766. Yet, even in this washout of a summer, temperatures were close to, or just below, the average. But then, this 'average' hasn't yet been adjusted to accommodate for what we have experienced or come to expect. Since 2000, records for temperature have been routinely smashed. We've simply got used to much better summers.

Better? Surely I shouldn't be talking in such terms if global warming is behind our recent bumper crop of summers. But it's hard for me to think otherwise, and I'm sure I'm not alone. Remember, 'warm and sunny' are the favourite weather conditions in this country, and sunny Spain the favourite holiday destination. When the UK media first took notice of global warming, in the early 90s, the headline grabber was the possibility that this might mean a more 'Mediterranean' climate for us in the future. Or maybe that's the only bit I remembered. In any case, I have to confess it is this hope that has secretly sustained me through the dark years, and stopped me emigrating years ago. The Mediterranean, with its glorious weather, relaxed lifestyle, café society, siestas, exciting food, beautiful people (need I go on?) is the land of our heart's desire, to which we regularly escape. And they tell us in the future our poor, damp, cloud-bedevilled island will become like this? Magically transformed into a version of holidayland, without the need to go anywhere? It's a wonder recycling has taken on to the extent it has in the UK.

Yes, I know. I should rise above my selfish demands for immediate gratification, and think long term and globally. I've been seduced by the selective headlines, and need to read the terms and conditions more closely. So, hoping to understand this issue much better, I arranged a meeting with Dr Peter Stott, of the University of Reading and the Met Office.

Dr Stott is an expert on the temperature of England and responsible for a breakthrough in the scientific debate on global warming. Stott and his colleagues have provided near-conclusive proof that an extreme weather event could be linked to human causes. They used mathematical modelling to prove the likelihood 'that past human influence has more than doubled the risk of European mean summer temperatures as hot as 2003'. They also projected that the likelihood of such events would 'increase 100-fold over the next four decades'.[3] By then the summer of 2003, which claimed thousands of lives across Europe, and which saw the highest temperature for England recorded, would be the norm or even seem mild. This was the man to terrify some sense into me.

The sun was shining brightly when I went to Reading to meet Dr Stott. It was mid-May, and the very end of the extraordinary spring of 2007. The University of Reading is set in a beautiful rural campus, and I passed numerous clumps of students on my way to the Meteorological Department. Youth was out in force that day enjoying nature in the glorious spring sunshine. They played frisbee, pretended to revise, or just lay on the grass indifferent to their doom. Their shouts floated through the open window as Dr Stott and I talked about global meltdown. He showed me the Met Office charts indicating just how anomalous the recent weather was. Like most scientific experts on the subject, he is cautious about attributing specific events to a global pattern (especially with our fickle climate); but the previous summer, winter and now spring were looking like clear markers on a steadily upward curve.

As I stood up to leave, I noticed how brown Dr Stott's arms were. I asked him if he liked the sun. He smiled, looked slightly shifty, as if deliberating whether to tell me something, and then shared the following anecdote. In February 2007 he had contributed to the Intergovernmental Panel on Climate Change conference in Paris. After the conference a reporter for BBC News 24 caught him on the hop. She stuck a microphone in his face and asked him whether any good might come of climate change for the UK. Impulsively, Dr Stott replied that it might mean we'd get some decent summer holidays. Though he quickly followed this up with a whole gamut of negative consequences, his initial response had been that of a Brit, not a scientist. He gave voice to what many viewers must have been secretly thinking themselves. Yes, I thought, and joined the students to make the most of the glorious spring sunshine. (Just in time – for the tide of summer turned that very evening.)

Climate change evokes a degree of ambivalence in many of us. Too remote, too unlikely, too good to be true? *The Rough Guide to Climate Change* surveys some of the prognoses for these isles over the coming decades, and concludes with the following statement:

> As for what to do in the meantime, Britons must be best off planning for a warmer climate while learning to live with a bit of uncertainty about exactly how warm it might be.[4]

Excuse me? We've been planning for a warmer climate for the last fifty years. Like some Home Guard battalion looking out for the Luftwaffe, we've been practising the drill since we first went abroad, and have been extra-vigilant since the phrase 'climate change' called us to arms. We stand to attention by our barbecues, and sit rigidly on café and pub terraces, eagerly anticipating the time when we can dispense with the heaters, windbreaks and fleeces and get a crack at the real action. Every time the sun shines and the temperature rises for more than three days together, the London papers sport a photo of metrosexuals enjoying the balmy night on a West End street in the wee small hours. The caption usually runs: 'Barcelona? No. Soho, 3.00 a.m. this morning'. The very fact that it's newsworthy – imagine the *Barcelona Bugle* trying to make news out of this: 'People spotted drinking outside till quite late. Weather believed to be the cause' – betrays how exceptional this is, and the pathetic desire we cherish for this to become the norm. The UK is the second largest market for convertible cars in Europe, with sales increasing threefold over the last decade. Oh, I think we're prepared.

A Londoner prepares for climate change
(Clapham Junction Station, April 2007)

As for learning to live with a bit of uncertainty. Is he kidding? Our middle name is uncertainty. How about a little certainty for once? From where I'm standing, climate change is starting to look more like climate striptease. A tantalizing glimpse of tanned torso, a lingering over-the-shoulder flash of sultry Spanish suggestion, and then boof! On with the winceyettes and windcheaters, and we're back in dreary old Blighty again. Mustn't grumble, we've had a good run lately. I'll put the kettle on. If this is Gaia wreaking her revenge, she's adopted a depressingly English incarnation to dispense our share. A stern governess who's withdrawn treats and ordered a cold bath for all of us. You've had your bit of fun, and now you will pay for it.

And to cap it all, there is the possibility that global warming might mean the slowing down or even the shutting off of the Gulf Stream. As you no doubt know, the Gulf Stream acts as a kind of conveyor belt pumping warm water past the western edge of Europe, and making things a darn sight milder than they should be at this latitude. But signs indicate this might be slowing down or even switching off, due to an influx of fresh water from melting icecaps. Which means it could be more like Moscow than Marseilles for us. UK weather could actually get worse. Now wouldn't that be just typical?

According to Kate Fox, 'typical' might be considered the English national catchphrase, reflecting the pessimism that is one of our defining characteristics: 'a delayed train or an undelivered dishwasher is "typical" in the same way that rain on a Bank Holiday picnic is "typical".' The Gulf Stream shutdown would be the mother of all Bank Holidays. Well, we didn't really expect climate change to be a picnic, did we? When it comes to rain, our glass is always a little over half-full. In Fox's view, we expect it to rain, and feel a gloomy satisfaction when it does, as it means we can trot out the other favourites, such as 'Mustn't grumble'. Mustn't grumble, as we know, usually terminates a truly marathon session of doing just that. We like moaning, and have turned it into an art form. If moaning is a quintessential English trait, moaning at something that can't be changed couldn't make us happier. We have a lot of weather. It's not generally the type we like, and there's nothing that can be done about it. Perfect. (What is this chapter, if not this book, but a protracted moan about the weather?)

If we believe Fox we wouldn't actually be very happy with a 'Mediterranean' climate, for then her 'moderation' rule would apply:

> even warmth and sunshine are only acceptable in moderation: too many consecutive hot, sunny days and it is customary to start fretting about drought, muttering about hose-pipe bans and reminding each other in doom-laden tones of the summer of 76.

And now, the summer of 2026 too. For we don't like 'too much of a good thing'. Whilst 'Every year, English holiday-makers, sighing at the prospect of "getting back to reality", comfort each other with the wise words: "but if it were like this all the time we wouldn't appreciate it".'[5] Holiday weather all the time, so you could plan things, live outside in the summer, come out of our shells and our castles to practise the lifestyle we've observed and essayed abroad? We wouldn't appreciate it? I'm not so sure.

But I was determined to find out. Encouraged by the spring of 2007, not to mention headlines such as that of the *Daily Express* for 16 April 2007: '80 degrees and . . . the good news is there's more to come', I'd intended to test Fox's thesis.[6] I was going to make careful observation and close and structured questioning of the English experience yet another heatwave, as a foretaste of what might be in store for us. Could the weather cure some of what Fox calls our 'social dis-ease'? Or would moderation and moaning come into their own? Do we expect rain because we are pessimistic, or are we pessimistic because we have learned to expect rain? I was prepared to find out.

But, of course, the weather had other plans. Typical. I was writing about something I hadn't seen for over two months, and it was seriously getting me down. If the Mediterranean wasn't coming to us this year then I would have to go to it. It is indeed impossible to live in a country that is permanently under hatches, with the rain thrumming down, so I did what poor Keats should have done much earlier. I bought a one-way ticket south, and brushed the dust and mud of England off my shoes for ever.

IN THE SUN

OK, not quite for ever. But long enough to answer a few important questions and provide some sort of conclusion to my quest. For wrapped up in my speculation about climate, identity and life lived under hatches is a bigger question; the one to which my investigation must ultimately lead, and one of the Big Questions that have occupied far greater minds since thought began. What is the source of happiness? Might sunshine be considered a candidate?

I can almost hear the snorts of derision from real philosophers, psychologists and economists whose job it is to ponder such questions, and who show little interest in such trivialities as the weather. I'm not talking in absolute terms here; perhaps no more than speculating whether more and dependable sunshine might dispel some of the gloomy negativity beclouding the British countenance. Every year it seems we read of a report revealing just how unhappy Britain is. A paper entitled 'The Politics of Happiness', published by the New Economics Foundation in 2003, summed it up: 'We are richer, healthier and longer-living than at any point in our history.' And yet, 'there is simply a sense among us, almost subconscious and rarely articulated, that life can and should be better than this – or at least, feel better'.[7] The NEF sets out to challenge orthodox economic definitions of progress, and to measure instead 'the things that really make us happy – and what gets measured matters'. This looked promising, so I asked a NEF spokesman: Does nice weather matter? Tackling climate change was certainly at the top of the NEF's agenda, he told me. But the weather? No comment. You can't build a political agenda on that.

Economists, old or new, might not pay much attention to the weather, but the economy does. According to Jim Dale, a 'risk meteorologist' and founder of British Weather Services, 'between 80 and 85 per cent of purchasing decisions are weather related. It's the biggest single factor affecting all sectors of the economy.'[8] It determines what we eat, what we wear, how we use our leisure time and where we are prepared to spend our precious holiday allocation. In a heatwave, absenteeism soars among indoor workers. In a washout, retail sales of many things plummet, and we retreat back under those hatches or flit off abroad. Lastminute.com reported record traffic on its website for the 'summer' of 2007, 'up more than 30 per cent on a year ago' at one point, whilst British Airways reported a 20 per cent rise in long-haul bookings on the same time the previous year.[9]

A few of these trippers might be wondering about ever coming back. In 2006 the Institute for Public Policy Research reported that 'Britain has more people living abroad than almost any other country'; that 'almost one in ten Britons now lives abroad and that a British national emigrates every three minutes'. Over the next five years another 5 million were predicted to follow suit. The main reasons for emigrating or considering this move were 'better lifestyle': 37 per cent, followed by 'better weather': 32 per cent. As Australia: 40 per cent and

Spain: 31 per cent were the favoured locations, this 'better' life is evidently found under bluer skies.[10]

Historical trends confirm this pattern. According to an academic study of Brits migrating to rural France, until the 50s and 60s 'the prevailing trend was of movement from southern Europe northwards'. These were classic economic migrants, moving from the south to the more affluent north in pursuit of employment opportunities. In the 70s this stalled, and from the 80s reversed, with northerners moving south, not for jobs, but for what these writers call 'consumption', meaning lifestyle. They cite an earlier study which found the 'primary reason why foreign residents were attracted to the Spanish province of Alicante was its climate, followed by its cheaper cost of living'. This is the main impetus for the new migratory trends, where holidaymakers wish to 'lengthen and broaden the consumption experience'.[11] In other words, and *contra* Fox, to make life 'like this all the time'.

This dream of permanence is evidently cherished by many. Hence the popularity of books like Peter Mayle's *A Year in Provence*, or Chris Stewart's *Driving over Lemons: An Optimist in Andalucia*. And hence the number of television programmes devoted to finding people *A Place in the Sun*, to quote the title of the most popular. *A Place in the Sun* has become a highly successful brand. In addition to the TV series, there is a magazine, with a circulation of 36,000 each month, and an exhibition which regularly packs out London's Excel Centre or Birmingham's NEC. It is the biggest overseas property exhibition in the world, attracting up to 21,000 visitors, with four shows a year.

Not all of these visitors are thinking of upping sticks permanently. According to Matt Havercroft, editor of the

A Place in the Sun magazine, 'the pensions collapse and the overblown UK property market have led more and more people to look to overseas property as an investment'. And yet the vast majority of buyers are still what he terms 'lifestyle purchasers', and 99 per cent of them are looking for sunshine, he reckons. The magazine simply can't feature a location or follow a happy purchase if the sun isn't shining. It is the estate agent's 'south facing' rule writ large. Both Matt Havercroft and Amanda Lamb, the TV show's presenter, put the success of the brand down to its name. It 'captures exactly the dream' of so many Brits, Amanda Lamb told me. As far as she knew, there was no equivalent TV show in our fellow sun-starved northern countries.

But have they found what they are looking for? Could blue skies and more dependable sunshine actually make us happier? There was only one way to find out. Spain is home to an estimated one million Brits, of whom 700,000 are believed to be permanent residents, with the majority on the Costas of the south. I needed to talk to them. I contacted Radio Europe Mediterraneo, the main English-speaking radio station for southern Spain. They kindly allowed me to broadcast a live request for volunteers to talk to me about their life in the sun. REM is based in Marbella, and the producer suggested a likely location in that town to meet people and conduct my interviews.

To supplement my fieldwork, I also posted a few questions on an ex-pat internet forum, and sent my questionnaire to friends and friends of friends who had lived in the sun for any length of time. There were also some lucky people I hadn't seen for years, who lived in sunny places, who just might welcome a house guest over the coming weeks. I would observe their lives at close quarters, interview them at length, talk to any other ex-pats they might know, and simply absorb the atmosphere of life under blue skies. I wasn't coming back to Britain until I could answer my questions. And I certainly wasn't coming back until the weather improved. So I sent out my plea across the Spanish airwaves, devised my question-naire, sharpened my pencils and blew the dust off my flip-flops.

A PESSIMIST IN ANDALUCIA

As the plane banked over the green fields of soggy Surrey, cut a path across sodden Sussex and headed south, it occurred to me that this was the first time I had ever bought a one-way ticket abroad. It gave my trip an exciting finality and purpose. And as we pierced the grey blanket that had sat on us for months, I revisited a question that has nagged at me for years: why on earth do I live in the green damp island disappearing beneath those clouds? Every bad summer I declare to be my last here, and press my face to estate agent windows abroad like a boy outside a bun shop. It's always struck me as a cruel mistake that I was born here, and I feel a wrenching home-sickness when I join the check-in queue back to the country specified on my passport. But then I would slip back into the groundhog day of perpetual grey, or a fine summer would miraculously pop up and reignite my secret hopes for a change in climate, and I give Britain one last chance.

I suppose it's all I know. I have roots here, friends and family, and something has always held me back. If I'm honest, it's the thought of loneliness, and the secret fear that sun-shine, long-cherished object of my desire, might not actually be the answer to everything. But this time I had passed the point of no return (fare). Driven away by the rain, with a one-way ticket and a project to find some answers, it was make or break time. I felt like Ratso in *Midnight Cowboy* finally catching his bus to Florida. Bad example: he dies just in sight of the palm trees. Keats, breaking free of those hatches: again too late. But maybe that's the point. I should recall their doomed bids for sunshine, and take action while I could. It was still only early July: there was time to ride into a sunset.

The fields next to the airport in Spain were baked golden; the sky impossibly blue. It looked as if it had never known a cloud, and never would. The temperature that greeted me felt like twice that I'd left behind an hour and a half before. It embraced me like a long-distant lover, and brought tears to my eyes. I had come home.

I had been to Malaga before, but had always taken a bus directly on to Granada. The bus I took today went in the opposite direction, along a strip-development megalopolis dedicated to sun-worship. This was the logical destination for my quest, and an obvious site for a pilgrimage. Yet my heart sank as the bus trundled along the coast.

It is customary at this point for the travel writer to express horror at the abomination that was now unfolding before me. To hold one's nose at the vulgarity, cheapness and ugliness of the hotels, bars and restaurants that crowd this – once charmingly pristine – stretch of coast now sacrificed to other nationalities', other classes' idea of fun. To shudder at the signs advertising fish and chips and full English breakfast, and the names of bars proudly declaring their allegiance with cheap little flags. To feign amazement at finding these things, and deploy the standard insult of 'Blackpool in the sun'.

Perhaps the most English thing about the Brits on the Costas is our response to them. Snobbery is acceptable, even obligatory, here. Giles Tremlett concludes his account of that twin-town excrescence on the Costa Blanca with the observation: 'I have never seen it, but I feel sure that somebody, somewhere is selling long, gooey, pink sticks of Benidorm Rock.' They already have 'the sort of entertainment once provided by working-men's clubs'.[12] In Blackpool itself, such things might evoke an affectionate tribute to traditional working-class hedonism. There it is 'folk'; here it is obscene: an unnatural growth bearing the colonizer's stamp of extraneous imposition – 'ghettos' of unintegrated Britishness making no concessions to local culture, traditions, materials or language. The Norman Tebbit test is applied here, and they are found wanting. I wouldn't expect the Spanish, Portuguese or Moroccans to stop selling their national delicacies in the cafés, shops and restaurants of Golborne Road in west London where I used to live. Or force them to use my language, eat my food or wave my flag. These places, and the locals who throng them, were why I loved living there. Sipping strong *café com leite* with Portuguese custard tarts in the Lisboa café, hearing nothing but Iberian languages, I could turn my back on the grey streets outside and pretend I was somewhere else.

Why is the reverse so very wrong? Yet somehow it is. My heart sank, because I saw what others had seen, and felt as they had felt. And I was trying so hard to keep an open mind. It's because this is supposed to be *somewhere else* that we are appalled at being confronted by what we have tried to leave behind. That need to escape again. Blackpool at least implies a certain jaunty exuberance, but parts of Torremolinos reminded me of the Elephant and Castle shopping centre. Beat that for an insult. Yet how could Benidorm or Torremolinos be anything but Blackpool in the sun? Resorts

like Blackpool grew up to cater for the leisure needs of a
working population. They were dedicated to fun, excess and
release, and fulfilled their function admirably.

What changed has already been documented. This same
population acquired means to pursue the same pleasures
with the addition of one extra ingredient that had become
essential for the annual holiday, but could not be relied on at
home. Supply met demand, in a hurry, and transformed the
fortunes of these regions. Unfortunately it also blighted them
aesthetically, environmentally and culturally. That's to say
nothing of the rampant corruption and criminality which the
solar gold rush ushered in. When the Spanish and Portuguese
sought El Dorado in the New World, one of the first things
these colonizers did was build churches like the ones they left
behind. These new temples to the sun, on the coast dedicated
to his worship, also bear the unmistakeable stamp of their
origins and purpose.

But the Englishness of our response runs deeper.
Blackpool simply has no right to be here. No right to spoil the
view of travel writers, but also no right to be enjoying itself in
the sun. Part of what defines the British holiday experience is
the solar lottery. The chance that it might just 'turn out nice
after all' – but probably won't. To transplant such trashy
Utopias to where the sun is guaranteed strikes at the heart
of English pessimism, puritanism and class-consciousness.
George Orwell was not a snob, or at least tried hard not to be.
Yet when in 1946 he read a proposal for 'the pleasure resort of
the future', with vast dance floors and landscaped lagoon
pools, he was appalled. These resorts would have sliding roofs
– 'for the British weather is unreliable' – and 'Sunlight lamps
over the pools to simulate high summer on days when the
roofs don't slide back to disclose a hot sun in a cloudless sky.

Rows of bunks on which people wearing sun-glasses and slips can lie and start a tan or deepen an existing one under a sunray lamp.' These artificial paradises captured Orwell's dystopian imagination, and differ radically from the scattered sunlit passages of his I noted earlier in this book; their slick contrivance and dishonesty provided further evidence of society's slide into barbarism. He concludes his essay by warning how many modern inventions, including cinema, the radio and the aeroplane, had the 'tendency to weaken [man's] consciousness, dull his curiosity, and, in general, drive him nearer to the animals'.[13] Or towards the sun.

Orwell was right to fear these things, as anyone who has beheld a planeload of passive, incurious, unconscious, music-blasting, night-clubbing, sun-absorbing Brits abroad can testify. And yet, bad though the package tourists are, there is worse. Giles Tremlett can at least allow them 'the pleasure of a two-week holiday away from the accounts department, tele-phone sales or factory floor'; but he simply can't explain or excuse the British residents of the Costas: 'Whereas those who install themselves in France's Dordogne or Italy's Tuscany often do so in a spirit of cultural inquisitiveness, these people seem to have been attracted only by sunshine.'[14] That is the greatest crime.

A Utopia based on sunshine would appear to be an impos-sibility. If it's true, as Wilde asserted, that the sun hates thought, thinkers are more than happy to return the compli-ment. J. G. Ballard's novel *Cocaine Nights* (1997) turns what Tremlett describes, and Orwell could only begin to imagine, into a dystopian fantasy. Set in and around the more exclusive part of the coast my bus was now entering, it shows what happens when the British move out to the sun, and devote their time to sun-lounger lethargy. They get bored of course.

These lotus-eaters in leisure suits withdraw into their apart-
ments for a diet of satellite TV, booze and prescription drugs:
'A billion balconies facing the sun . . . The Costa del Sol is
the longest afternoon in the world, and they've decided to
sleep through it.'[15] The sun that drew them here exerts a sinister force in Ballard's fantasy, trepanning them into the kind
of animal stupefaction Orwell foresaw. Too much of a good
thing. It takes an engineered outbreak of petty crime and
racier narcotics to shake up their sun-sanitized world, and
restore life into the leisured dead. Very Wildean. A taste of the
gutter to make them appreciate the stars.

At the other extreme there is *Eldorado*. This soap opera,
which ran for only a year until it was axed in 1993, was set near
Fuengirola among ex-pats. Like most English soap operas it
featured a cast of working-class characters, this time in the
sun. Why did it fail? Partly because English soap operas are
heirs to the Kitchen Sink tradition, and turn the national
pastime of moaning into a spectator sport. Their cathartic
function is to demonstrate that life could actually be worse.
No one should want to live in one. But many of its target audience would like to live in southern Spain. So it could either
violate the rules of the genre by showing people contented
(which would be boring), or shatter this dream by introducing
the usual quota of misery. Which it implausibly tried to do.
When characters escape Albert Square or The Street to the
sun, we're not supposed to follow them there. Misery begins
and should remain at home. It is the dream of escaping to the
sun that sustains so many people, the cherished dream of
making holiday every day. The audiences were just not prepared to have this golden key snatched away from them.

So who is right? Ballard or the (non-)viewers of *Eldorado*?
The next day I hoped to find out. I arrived in Marbella just as

the sun sank into the sea, and was surprised by what I found.
This was more like Bournemouth than Blackpool – or perhaps
'Ibiza for old people', as a friend more cruelly put it. The old
town, where I was staying, does a fair impression of an
Andalucian village, with picturesque plazas and bougainvillea
tumbling over crumbling walls. It also had what looked
like the 'authentic' tapas bars and bodegas frequented by
Spaniards that I usually seek out when I'm in Spain. But this
time I was in search of their opposite. On my first night,
however, I completely failed to find them. The café where I
would be receiving volunteers the next day was full of Spanish
people, and nothing like what I had expected. The old town
was quaint, the beach front 'classy', and I realized I'd landed
in the showpiece of the Costa del Sol. I went to bed wondering if I'd have to go back along the coast to Blackpool to find
what I was looking for.

The next morning I set up my vigil at the Cafeteria
Marbella as advertised on REM, in the last patch of sunshine
on the terrace. I ordered a coffee, and adopted a 'writerly'
demeanour complete with notepad and welcoming expression. But few of those I scrutinized fitted the picture I had
formed of a resident Brit, and nobody approached me. After
ten minutes there was a whirring sound. I looked up to see the
sun disappearing behind a mechanized shutter. I was not
having a good morning. Another coffee, another hour, and still
no punters. There was no way I was sitting here all day in the
shade. So I abandoned my post to go in search of the Brits.

I had the telephone number of the British Association of
Marbella, which sounded promising. It was answered by a
very nice woman called Mary, to whom I explained my
request. She told me some of her members would be playing
Scrabble that afternoon at the Lucky Leprechaun, in a part of

town I hadn't yet been to. They'd be happy to talk to me. They were. My luck changed with the Leprechaun, and I found exactly what I was looking for. So, it would appear, had those whom I met over the next few days.

I soon worked out why my public broadcast had not brought a stampede of voluble volunteers. People, especially the British, are generally most vocal and forthcoming when they have something to moan about. The people I met, every single one of them, were very happy here. They have their troubles, of course. Work, for the non-Spanish, can be tricky to find. Partners die, people get ill, pensions don't stretch as far as imagined. Life and its problems don't stop because you've jetted off to the sun. Being an alien often exacerbates them. But what I did witness was a strong support network. The climate encouraged life outside, and that meant socializing. Socializing makes the kind of isolation someone might experience in cold, keep-ourselves-to-ourselves, don't-like-to-pry Britain less likely. What they had here was 'community'. Coming from London I know just how precious that is.

The elderly claimed the climate kept them young. 'I'd be dead by now if I'd stayed in Scotland,' said one woman. It kept them healthy, active, sociable, alive. Those I met were living proof of this. They appeared spry and colourful, and looked much younger than their years. The regulation M&S drab of their counterparts back home, or the Saganauts passing through, was nowhere to be seen.

The magic didn't just work with the elderly. I met Paul in a group of long-term residents sitting outside what advertised itself as an 'English bar' I stumbled across one night near the old town. He was, I'd say, my age, and had lived here all his adult life. He left his home town of Leicester when Thatcher was still in power. 'Better skint in the sun than back home,' he declared. He had lost a kidney a few years back in a bicycle accident, his small real-estate business had dried up when the big competition moved in. But he was happy and looked forward to business picking up soon, as he was certain it would. This was the original optimist in Andalucia. I tried to imagine such a conversation in a pub in Leicester. No work; an accident. Actually, I couldn't imagine walking into a bar in Leicester or anywhere in England, and talking to complete strangers so easily. Nor could I imagine what I witnessed at the end of the evening. The working-class men who left the table hugged each other goodnight. Unselfconsciously, without embarrassment, cheap jokes or some other macho ballast to ensure no one misunderstood their gesture. And they weren't even drunk. English bar? I think not.

So many other supposed English traits had been left behind with the rain at Gatwick. Whilst I was awkward and reserved, everyone else was friendly and welcoming. I asked to be introduced. Introduce yourself; they won't bite. Moaning? The only one moaning was me. Principally about the weather back home. When they did moan it was about the country they had left behind. Many long-term residents (up to thirty-five years in some cases) were happy to renounce their nationalities. Pie and chips may have been on the menu, but I got the impression that was mostly for the more timid tourists, or occasionally for the residents when they feel homesick. The ones I spoke to mentioned Mediterranean food as part of the better 'lifestyle' they enjoyed here. As for the British Association, I met all nationalities, brought together simply by an ability to speak English.

And the weather? They were in no hurry to talk about it. I had to keep bringing them back to the subject. What about it? 'It's wonderful, what do you think?' But how? What difference does it make? 'All the difference in the world. You live outdoors. You really live life to the full.' Winter is two mild months, with a bit of rain. And what of sunshine, the thing that had brought many of them here in the first place? 'I love to wake up every morning and see the sun. Every day is brighter . . . But after a while you get used to it. Tend to take it for granted'. . . 'I loved it when I was young, but can't really take it now.' The longest-serving residents were proud to declare they avoided it (like real locals). 'I haven't been on the beach for twenty years,' boasted Luigi (the Welsh-Italian owner of an Irish bar living in Spain. How's that for diversity?). Many longed to escape in July and August when it gets too hot and crowded with sun-seekers. A holiday from the sun. Truly the grass is always greener. And yet no one felt nostalgic for British weather. Even if it hadn't brought them out there, it prevented many of them from returning. The long-termers simply couldn't imagine spending a winter back in Britain.

My harping on about the sunshine marked me out as a tourist. As did my need to get back into it. The bars were cool and dark, and I could only spend so long inside before I cracked. 'Do you mind if we sit outside to continue our talk,' I asked Frank, one of the longest-serving residents. So we moved. He to the shade, me to the sun. Frank had come here in 1962, on his way to Morocco, but decided to stay in Marbella. There were only about thirty foreigners back then, he claimed. He opened a beach bar, played the Beatles to the locals, and built an empire. According to Frank, the Costa del Sol was no longer real Spain, but Europe's holiday playground. 'We're all Europeans now, and this is Europe's south coast.' It was one big melting pot. The sun did all the melting.

THE WINDOW

From Spain, I flew to southern France, to stay with friends and get another perspective on my question. According to the 1994 study of Brits migrating to France, the British made up the highest concentration of EU non-nationals in the country. They were mostly middle class, sought rural locations rather than the coast, were dispersed across a number of regions, but with a high concentration in the south. Although better weather was a major factor for many, there were other motivations for relocating or buying a second home in France. Many claimed to be 'Francophiles', and were seeking to immerse themselves completely in their adopted country. France, therefore, offered a good contrast to the received view of the Brits in southern Spain.[16]

I'm not aware of any dystopias written about British people moving to rural France to live the full-rich life. The opposite view, best represented by Peter Mayle's popular *Year in Provence* (1989) and its sequels, is much more prevalent. Mayle and his wife had been there before as tourists, 'desperate for our annual ration of two or three weeks of true heat and sharp light. Always when we left, with peeling noses and regret, we promised ourselves that one day we would live here . . .' That day came, and his books chart his gradual integration into Provençal life and identity.

The aim of the game is to lose your Englishness. By May of his first year 'we realized we were becoming as obsessive about food as the French'.[17] Ten points. After a drought, he even prays for rain: 'a promising sign that we are becoming less English'.[18] Twenty at least. His integration appears to be complete once he is despising, or is unable to recognize, his former self:

a funny bunch, these natives of August. It was impossible to miss them. They had clean shoes and indoor skins . . . The beauties of nature were loudly praised . . . and I particularly liked the comments of an elderly English couple as they stood looking out over the valley. 'What a marvellous sunset,' she said. 'Yes, 'replied her husband. 'Most impressive for such a small village.'[19]

No doubt he gave a perfect Gallic shrug, had another truffle butty, and rolled his eyes to those spectacular heavens he now took for granted, but had no intention of sharing.

As George Bernard Shaw once said, 'It is impossible for an Englishman to open his mouth without making some other Englishman hate him or despise him.'[20] By resenting the incursion of *parvenu* compatriots, Mayle hoists a flag of St George as high as the one adorning the sweaty football strip of his much denigrated counterpart on the Costa. Whilst Costa-man is happy to share his good fortune with his fellow expatriates, enjoys a gregarious life socializing with them, and clasps them close to his bosom, Mayle-man wishes to pull up the drawbridge on his chateau, maintain his moat, and enjoy his little island of integration alone. As his French neighbours happily accept this foreigner into their community, Mayle must ultimately lose his integration points for refusing to do the same. The 1994 study quotes a characteristic view from one of its Brits in France: 'The French should impose an embargo on people from foreign countries in areas of concentration like Dordogne.'[21] Loire-di-dah.

It's not their fault. It's simply how we are programmed. It is merely a version of the British beach rule. When a beach fills with Brits it obeys a centripetal movement, starting with the peripheries and gradually filling in. This is a small crowded island, and we want our space, and that means being as far away from other people as possible. So the first to arrive go to the furthest remove, trudging away from the path to ensure solitude. The next arrivers (who now constitute 'intruders') will maintain the furthest distance from these existing annoyances, and so on. Through grudging, scattered centripetal infill, the beach will eventually be 'ruined' by a mass of people all wanting the same thing. To enjoy the beach alone. The Spanish, Italians and southern French seek the opposite when they come to the beach. They are looking for fun and company, not solitude. They come there for life, and so cluster together with strangers. Their beach fills up centrifugally, radiating outwards from the vibrant centres of life.

A similar social physics can be applied to the differential distribution of Brits in France and Spain. Whilst Costa-man clusters to the centres of life like his Spanish hosts, Mayleman retains his ingrained desire to escape our crowded little island, and enjoy his rural retreat in splendid isolation. At least isolated from those he crossed the Channel to avoid. It would appear that there's only so much the sun can achieve, and you don't have to eat fish and chips abroad to display your true national colours.

When we touched down at Toulouse the fields were green, the skies grey, my mood black. The TV on the bus that took me to the city centre was showing the weather. The whole of northern France was besieged by rainy symbols steadily pushing south. People around me discussed the weather. In my insularity, inverted patriotism and pessimism, I'd assumed that Britain had been suffering alone. Turns out the whole of northern Europe had shared much of our misery this year. They had not known a summer like it in Toulouse. Some expressed relief, as it could reach 40 degrees plus in the summer. An old lady involuntarily fanned herself with her paper as she remembered. This was cold comfort to me, and clouded the prospects for the rest of my holiday – er, research trip.

I was taking a train further south the next morning, so I might escape. Down to the Mediterranean, near the Spanish border, where my friends had just bought a lovely apartment. The clouds stuck with the train as far as Perpignan, attached to its top like a kite surfer's kite; but the cord was eventually cut, and by the time we reached the coast, sky mirrored sea in perfect blue. Phew.

Helen and Phil had lived in France for thirty years, in Toulouse mostly, but with a lovely ramshackle cottage in a remote village in the Tarn valley, which Phil had restored from a ruin; now they had this weekend apartment overlooking the sea. This was their first week in their new place, and it was exciting to discover the town with them. They have two teenage boys, Louis and Ben, who have never lived in Britain. They had been on a visit a few weeks before, however, and their experience of two weeks of Welsh rain and the need for hot water bottles in June made it easier for them to under-stand why I was writing this book. Yet most of the time I was there the boys spent indoors, playing some accursed com-puter game, or with their noses stuck in the final Harry Potter book that had just arrived. They had no interest in the sun, and took it completely for granted. Helen liked it, yet took it for granted also. She topped up on sunshine simply by walking about, and rarely bothered to sunbathe these days. This was pretty much the view of most people I consulted over the coming weeks. I had the deepest tan at every table. The true mark of an outsider, in every sense.

My hosts' indoorness caused me something of a dilemma.

I was torn between the polite obligations of a guest and the need to be out in the sun. I felt like a secret tippler. Sloping off for a crafty top-up, volunteering for errands that would take me outside, standing in the open window to chat when the sun came round so I could have the best of both worlds. The sad isolation of an addict. I started to take those scientific papers about UV addiction a little more seriously. I'd been in constant sunshine for over three weeks now. (After Marbella I'd spent some time in Madrid, staying with an old school friend Danny, a fellow sun-obsessive with whom I used to have tan-offs. He'd lived eleven years in Spain, and my white bits were browner than the rest of him. But as he smugly remarked, it was only July, his summer had only just started.) And still I had an intense physical craving for the stuff. Seeing how such behaviour went against the grain of southern life, I began finally to accept the addiction hypothesis. I wondered how I might fare myself. How long would it take before I began to seek the shade? I wondered if I ever would. From observing others I started to observe myself.

It was six o'clock one afternoon, a few days into my visit. We had just finished a late and elongated lunch (something I was starting to get used to), and dispatched a few bottles of the local rosé. Standing within the shaded depths of the apartment, I noticed the open window beside which I usually loitered to catch the sun. Wooden shutters, peeling white paint, fringed by gauzy white nets flouncing in the gentle breeze. I'd stood by it often, urgent for my fix, but now I looked at it properly for the first time. It had silently arranged itself into a painting. The background (about a third of the composition) was deepest blue, shared between sky and sea. The bottle-green cliffs beyond took a scoop out of this blue, creating a dramatic diagonal that was completed by the crescent of grey pebble beach in the foreground. This was now flecked with painterly splodges of bright beachwear, as the bathers returned for the kinder evening sun.

True colour returned and sang out with the bathers, unmixed from the sun's palette. The rooftops were glazed with butter, the sea was set a-dance with diamonds. All his own work. The sun itself had withdrawn from the picture he had lit, coloured and arranged so impressively, and was no longer falling through the window where I would bask. To get back into the sunshine, I needed to descend some steps, and in two minutes I could enter the composition. But I didn't. I remained on this threshold, out of the sun yet with an intense appreciation for what it had created. A phrase came to me, the herald of a new understanding: to possess, not be possessed by the sun.

When Van Gogh came south he was in search of a 'stronger sun', and was pulled towards it like one of the yellow flowers he painted so obsessively. He was perhaps the first artist to paint the sun itself, stripped of allegory or personification. Even Turner, self-pronounced sun-worshipper, swathed his god in mists or trailing clouds of sublimity; approached it with the due reverence of poetic allusion, yet never quite rent the veil of its temple aside. The Impressionists painted what the sun did; Van Gogh what it was, or how it felt. He looked it straight in the eye, and immersed both himself and his art in it: 'the sun in these parts, *that* is something different . . . I feel it is excellent for me to work in the open air during the hottest part of the day . . . I enjoy it like a cicada.'[22] His canvases offer no escape from the terrible beauty of what he saw and absorbed. This sun dominates a painting like *The Sower* (of which there are several versions, all with vast solar orbs). Occupying the picture's central vanishing point, it is the blazing heart from which the whole composition, expressed as colour, radiates. The pulsing strokes of molten yellow and violet pay homage to the true germinator of colour and life, and, perhaps for the first time in art, render the *sensation* of its fierce intensity. Yellow House, green fairy, ultraviolet, infrared – the very palate of madness: 'Oh! That beautiful midsummer sun here. It beats down on one's head, and I haven't the slightest doubt that it makes one crazy.'[23] Poor Vincent was drawn into what he drew. The artist as Icarus. Possessed by his subject.

The artist as Icarus. Drawn into what he drew (Vincent Van Gogh, *View of Les Saintes-Maries*, 1888)

To possess, not be possessed by sunshine. To know it is there, getting on with its work, allows you to get on with yours. This is surely the secret of living with the sun. Can I confess, I have occasionally found the presence of the sun something of a nuisance? The need to make the most of it means I have a constitutional inability to do anything else when it is out. Even in a good summer I can't quite bring myself to trust it won't go away as soon as I take my eye off it, so I don't see any films or plays; all home improvements are put off till winter. The sun, when it is here, possesses all my time. Possesses me, in truth.

And then it happened. One morning I woke up in Toulouse (we'd been there about a week under clear skies), and it was completely overcast and threatening rain. But what I felt was relief. For it meant I could get some work done for the first time in ages. I had been reading, making notes and thinking, but mostly chatting, eating and sitting in the sun. I'd slid imperceptibly into holiday – into the slow honeyed drip of southern light and time. Every evening the BBC news broadcast scenes of England under water, so I was completely aware of my good fortune to be here. But I was also aware of a deadline, and needed to get back to work. I had already sac-rificed one laptop on the solar altar back in April when it over-heated through repeated use outside, and feared doing this again with its replacement. I worked hard that day, writing up the Spanish leg of my journey. The clouds cleared that after-noon, but I remained inside working, shutting out the siren call of the sun. I felt I'd achieved something.

The scenes of floods also brought another understanding. As July slipped into August, with still no sign of a let-up, it became clear that this was no ordinary bad summer, and we were not suffering alone. This was truly a summer without hope of reprieve. As our summers were not usually *this* bad, I conceded that they were not usually *that* bad after all. This appalling summer conferred a kind of posthumous forgiveness on the ones I had cursed before. I looked back and saw their rather mixed characters in a different light. As childish caprice, rather than malicious intent. After a while I realized I had reached a sense of equanimity, a softening of my iron demand of the heavens. Slowly there crept upon me a hitherto unknown sense of solar saturation, even surfeit. With it came an ability to view what I had heard, seen, felt and understood with more philosophical eyes. It was time to conclude my quest, and assess what I had discovered.

9

RETURN

Thus, though we cannot make our Sun
Stand still, yet we will make him run.
Andrew Marvell, 'To his Coy Mistress'

So I had come full circle: finding myself at the end of another appalling summer, hoping to understand the desperate need for sunshine. What had I learned? Is it the answer to happiness?

Clearly everyone who possesses abundant sunshine appears to take it for granted. The reverse also holds true. The southerners I met living in England this summer were going as crazy as me. They finally understood our behaviour when the sun came out, came to appreciate what they had left behind, and some were seriously considering returning to it. Why on earth would you *choose* to live here, I demanded. Why on earth do you choose to remain, they sensibly countered. Yet it appears to be a quirk of human psychology to want what you don't have. Especially in our culture, where 'want', which originally meant lack, now principally means 'desire'. We desire precisely what we lack.

Psychologists call it the 'hedonic treadmill'. This metaphor explains why people expend a lot of energy pursuing goals they believe will make them happier, but which, when they gain them, fail to secure a lasting increase in their well-being. They then pursue further goals – an even better job, larger house, sexier spouse, flashier car, dandier suit, holiday home in the sun – in the misguided belief that happiness is just over that horizon if they only run fast enough to reach it. A treadmill, because no matter how hard you run to reach these goals, you remain at around the same base level of happiness.

This circumstance has been used to explain why the richest nations are not necessarily the happiest. Once wealth raises a society above a certain threshold of want (lack), there is a law of diminishing returns on what is acquired through its wants (desires). It is such gaps that have led thinkers to call for a new non-materialistic economy of happiness, and I naively proposed sunshine as an answer. Yet the available research doesn't support my suggestion. When global indexes of happiness are compiled, there is no obvious correlation between sunny skies and sunny dispositions. A recent survey found Denmark and Switzerland to be the happiest nations, closely followed by Austria and Iceland. The first obvious blue sky contender was the Bahamas at fifth position: a place it shared with Finland and Sweden. The favourite countries for British migration, Australia and Spain, were twenty-sixth and forty-sixth in the rankings. The UK, at forty-one, was actually higher than Spain. It would appear that attitude, ultimately, had little to do with latitude.[1]

A paper published in the journal *Psychological Science* in 1998 specifically tackled the belief that greater happiness is found where the sun shines.[2] It compared the reported levels of happiness of students living in California with those in

Midwestern universities, and found no significant difference. And yet the Midwesterners assumed the Californians would be much happier, largely due to their climate. The Californians rated their own climate higher than the Midwesterners did theirs. They also had a higher estimation of their own climate than of the climate of the Midwest. On the weather issue everyone agreed. Yet, it was a relatively minor factor for the Californians, and made no major impact on their overall levels of happiness. Whilst they took their climate for granted, it was a defining factor in the blissful picture the Midwesterners had formed about what life must be like in California. In short, climate dominated Californian dreaming, but not the reality.

The researchers called this phenomenon a 'focusing illusion'. It also explains why, after a surprisingly short time, both lottery winners and paraplegics become no more or less happy than they were before their circumstances changed, despite what both had imagined before. They adapt. We all adapt. It's what drives the treadmill. Better weather, vast riches, and even major disability are but parts of a bigger picture. When you focus on them exclusively from afar, you exaggerate their impact on your life. When you get there – rich, suntanned or even disabled – this circumstance falls into place, and you simply get on with life. Other desires, dreams, fears, worries and memories start to occupy and preoccupy you.[3]

So in theory sunshine cannot be the answer, because there cannot be one single answer. Whatever you set your heart on having is probably not the answer. Not that row of lottery numbers just waiting to fall into place, nor the sun lounger permanently reserved for you by your pool, nor that apple hanging temptingly on the tree. We're back in that Garden again, where we once had perfect weather all year round, but blew it with our itchy fingers and nagging desires. The fruit we had just couldn't be as tasty as the one we were denied.

I didn't need to fly to Spain and France to understand that. But I may have found out the hard way. When commentators explain what makes countries like Denmark happy, factors like 'close ties to friends and family' are ranked the most important. And socioeconomic factors are more significant than climatic. Another study, published in 2000, which also ranked Denmark top of the happiness charts, found the unhappiest countries to be 'Portugal, Greece and Italy, which are sunny'. The author remarks how Scandinavian communities are 'small and homogeneous. People there have a similar world view and a similar religion, so that it's easier for them to communicate and to understand each other's motives.'[4] Emigration, in any direction, severs those roots. The fragile plant of happiness might not thrive in foreign soil, no matter how brightly the sun shines over it. Believing there is a key can make anywhere a prison. The human heart can make a hell of heaven or a heaven of hell.

Happiness, counsel the wise, comes with renouncing desires. In his poem *The Nature of Things*, the Roman poet Lucretius contemplated those toiling on the ancient equivalent of the hedonic treadmill:

> striving night and day with surpassing effort to struggle up the summit of power and be masters of the world. O miserable minds of men! O blinded beasts! In what darkness . . . is passed this term of life . . .

Stand at the top of Liverpool Street station escalator at 8.30 on a weekday morning and you are sure to recognize this picture. City boys and girls clambering upwards to toil away their brief day in darkness in pursuit of gold and power.

But, purveyor of the Epicurean ideal, Lucretius proposes a simple alternative to the vanities he sees possessing his fellow citizens. Let the many strive to acquire sumptuous palaces decked with gold, and enjoy costly banquets; yet, these things are wholly superfluous to what the body needs, and where true happiness can be found. This is found, he claimed, by those who sit

> on the soft grass beside a stream of water under the boughs
> of a high tree and at no great cost pleasantly refresh their
> bodies, above all when the weather smiles and the seasons
> of the year besprinkle the green grass with flowers.[5]

OK, I'll buy it. I'll reject worldly vanities, and seek nothing more than to sit by a stream with a simple repast while the weather smiles. I'll arrange a picnic. But . . . cue thunder . . . Typical. In Rome, you could take such simple pleasures for granted, and therefore propose them as an antidote to the vanities of material gain. Rome has approximately eighty-one days of sunshine each year, compared with our paltry forty-eight. What if you subscribe entirely to Epicurus's doctrine and wish to join his school (called 'The Garden'), and yet by fate's cruel caprice you are born in a suburb of Croydon? What do you do? Run towards your desire, only to discover it was a delusion like all the rest? Renounce the desire, simply because it is your desire, and live in a darkened cave on bread and water? Or consume yourself with frustration believing there is a key whilst remaining locked in the prison? Or maybe there is a third option, and you don't have to travel any-where or do anything to achieve it.

In 1819 the weather curse of the previous three summers finally lifted, and Keats rejoiced. On 28 August he wrote to his sister: 'The delightful Weather we have had for two Months is the highest gratification I could receive – no chill'd red noses – no shivering – but fair Atmosphere to think in . . .'[6] And think, and write, he did. His 'Ode to Autumn', which many consider his most perfect lyric poem, was written a few weeks later. This is a poem, so Jonathan Bate claims, about and shaped by the weather; both the three lousy summers that preceded it, and this 'highest gratification' when it changed.[7] A celebration of the maturing sun, who sets

> . . . budding more,
> And still more, later flowers for the bees,
> Until they think warm days will never cease,
> For Summer has o'er-brimm'd their clammy cells.

The poet is acutely aware they will cease, and may never be seen again. He appears to be on a mission to experience the season, *as it's supposed to be*. To write the definitive autumnal description, and exult in what has been denied him those years before.

> Sometimes whoever seeks abroad may find
> Thee sitting careless on a granary floor,
> Thy hair soft-lifted by the winnowing wind;
> Or on a half-reaped furrow sound asleep,
> Drowsed with the fume of poppies, while thy hook
> Spares the next swath and all its twinèd flowers . . .[8]

Time, drowsed by the sun, stays his scythe but momentarily, and lets us glory in his bounties. This is perhaps the consummate expression of the spirit of making the most of it. The fullest flowering of that extraordinary year, it was also the last complete poem Keats ever wrote. A thing of beauty fashioned by the vagaries of our climate, and for this much we should be eternally grateful.

Our uneven allocation over the years has instilled a desperate desire to make up for lost time; for lost sun. Our desperate bingeing is no doubt rather tragic. But isn't everything that points to the brevity of life, of time slipping through our fingers, like the sand we sieve dreamily in our beach-side devotions? Scarcity and uncertainty have been motifs running through my story. They point to what is perhaps its ultimate theme: Time.

Sunshine might be made up of wavelengths, experienced as chemicals or displayed through skin pigment, but its ultimate measure, its ultimate value is temporal. From the heliotherapists carefully timing the exposure of their patients; to the fashionable displaying their tans that could only have been perfected over time and therefore overseas; to love's brief season finding its perfect metaphor in a summer's day; to the tricks played by time in our remembrance of lost sun; to the dream of permanence that impels so many from the north to secure their place in the sun – the sun casts its shadow upon a dial that counts our hours and days. Louis MacNeice's poem 'The Sunlight on the Garden' (1937) offers the definitive statement on this relationship:

> The sunlight on the garden
> Hardens and grows cold,
> We cannot cage the minute
> Within its nets of gold,
> When all is told,
> We cannot beg for pardon.[9]

What do you give to the person who has everything? The only thing he or she can't have. More time. The next best thing is the awareness of this fact. Sunshine provides this. It is because the nets of gold break, spun as they are with the fragile threads of time, and from an uncertain spool, that makes us grateful for those moments of sunlight on the garden. We are afforded an Epicurean ability to enjoy the most simple pleasures. To sit by that stream or in that garden when the weather finally smiles. By never taking it for granted, we are given, fleetingly, a taste of life's full sweetness and its awful brevity. For Time is all we have.

Sunshine can define a moment, make an occasion, preserve and sanctify a memory, and transform the merest patch of ground into a corner of paradise. It can only do this because it is not there all the time. And on this short day of frost and sun, we are dementedly determined to squeeze the last drops out of it. To experience those moments simply for those moments' sake. Making the most of 'it' is really making the most of life.

Golden halls might not bring happiness, but halls, meadows or gardens gilded momentarily by the glory of the sun surely can. And for the same reason. We value sunshine because it is scarce, and industries have sprung up to exploit these desires. And so sunshine has been turned into a commodity, to be consumed by the hour or sold by the square metre when we seek to secure it in abundance. A sunlit garden is waiting for us, at a price.

Yet gold and gems are valued not just for their rarity, but their lustre, the glimmering reflection of the distant star of which they are but shards. The alchemists understood this. As gold was the most precious metal, so the sun was its symbol for the alchemists – a circle with a dot in it. The sun is alchemical still. Transforming the dross of mundane existence with 'colour glorious and effect so rare', in Milton's words.[10] Creating pools of pure enchantment wherever its precious rays collect, and performing a synaesthetic symphony in the grey auditorium of our lives.

The alchemists' endeavour to turn lead into gold was not just for material gain. They sought to bring the metal to the state of perfection to which they believed all matter must aspire – an allegory of the soul's journey towards enlightenment. What they ultimately sought was the Elixir of Life, 'the Flower of the Sun'. As Ben Jonson asserted in his play, *The Alchemist* (1610):

> He that has once the Flower of the Sun,
> The perfect ruby, which we call elixir . . .
> Can confer honour, love, respect, long life,
> Give safety, valour, yea, and victory.[11]

That flower turns to the sun, and so do we. For sunshine is what it always has been, the very thing I first considered and then rejected because it was too elementary, too complex and too large to comprehend. It is simply the source of all life as we know it. We are but dust from that golden star. And we've got to get back to that garden.

Humankind was born on the plains of Africa, under sub-equatorial sun. We were surely not meant to wander this far north. I know I wasn't.

It's time I went home . . .

NOTES

1. PRODIGAL SUN

1. Graham Reynolds, *Turner* (London: Thames & Hudson, 1969), 12.
2. John Ruskin, *Fors Clavigera: Letters to the Workmen and Labourers of Great Britain*, vol. 4, letter 45, ed. Dinah Birch (Edinburgh: Edinburgh University Press, 2000), 194.
3. J. Wykeham Archer, 'Joseph Mallord William Turner, RA', *Once a Week*, 6 (1 February 1862), 166.

2. NORTHERN SKY

1. Jacquetta Hawkes, *Man and the Sun* (London: Cresset Press, 1962), 58, 250, 122.
2. 'The Sun-Dial' was a typescript publication circulated among the private members of the Sun and Air Bathing Group, run by Hugh Shayler, who met at various camping sites around the south-east in the early 30s. The copy from which this quotation was taken is held in the archive of British Naturism, and I am very grateful to Michael Farrar, their archivist, for the happy and productive days I spent with these papers.
3. According to Ronald Hutton, 'early Christian tradition' did not identify any date for the birthday of Jesus Christ. He quotes an early Christian writer, Scriptor Syrus, on why this

date was adopted: 'It was a custom of the pagans to celebrate on the same 25 December the birthday of the Sun, at which they kindled lights in token of festivity. In these solemnities and revelries the Christians also took part. Accordingly when the doctors of the Church perceived that the Christians had a leaning to this festival, they took counsel and resolved that the true Nativity should be solemnized on this day': *The Stations of the Sun: A History of the Ritual Year in Britain* (Oxford: Oxford University Press, 1996), 1. The festival coincides with the winter solstice. This is when the sun is at his weakest, and the celebration at this time was partly to encourage his returning strength after the shortest day has passed. When summer holiday adverts and brochures start appearing immediately after Christmas we might see this as a reassertion of the solar cult to which this season was originally dedicated.

4. David Whitehouse, *The Sun: A Biography* (London: Jon Wiley, 2005), 41, 43.

5. Oscar Wilde, 'The Decay of Lying' (1889), in *The Soul of Man Under Socialism and Selected Critical Prose*, ed. Linda Dowling (London: Penguin, 2001), 189.

6. Philip Eden, *The Daily Telegraph Book of the Weather: Past and Future Climate Changes Explained* (London: Continuum, 2003), 10, 12.

7. Antony Woodward and Robert Penn, *The Wrong Kind of Snow: The Complete Daily Companion to the British Weather* (London: Hodder, 2007).

8. Samuel Johnson, from *The Idler*, no. 11, 24 June 1758, in *The Idler and The Adventurer*, ed. W. J. Bate et al. (New Haven: Yale University Press, 1963), 37.

9. It is certainly true that the US has a TV channel entirely dedicated to the weather, and so might seem more obsessed. Yet it is a big country, subject to extreme weather conditions that people critically need to know about. The US also has a lot of channels – for just about everything, it would appear.

10. The survey, which was conducted by H. Neuberger in 1970, involved 12,000 paintings from British, American, Low Countries, French, German, Italian and Spanish artists. The average percentage of relative cloudiness was 34, Italy measured 22, the British school managed 48 per cent cloudiness and 0 per cent clear skies: cited by John E. Thornes, *John Constable's Skies* (Birmingham: Birmingham University Press, 1999), 29.

11. J.-K. Huysmans, *Against Nature*, trans. Robert Baldick (Harmondsworth: Penguin, 1959), 133–4.

12. Oscar Wilde, op. cit., 184.

13. Robert Louis Stevenson, *The Strange Case of Doctor Jekyll and Mr Hyde and Other Tales of Terror*, ed. Robert Mighall (London: Penguin, 2002), 23.

14. Peter Ackroyd, *Albion: The Origins of the English Imagination* (London: Chatto & Windus, 2002), 73.

15. Bruce Robinson, quoted in a documentary about the film included in the DVD.

16. Jeremy Paxman, *The English: Portrait of a People* (London: Penguin, 1999), 124.

17. George Orwell, *Nineteen Eighty-Four* (London: Secker & Warburg, 1949), 119, 139, 222.

18. Ruth C. Engs, 'Do Traditional Western European Drinking Practices Have Origins in Antiquity?', *Addiction Research*, 2/3 (1995), 227–39, 229, 228.

3. HEALTH

1. A. Rollier, *Heliotherapy: With Special Consideration of Surgical Tuberculosis* (1915; Oxford: Oxford University Press, 1927), 40, 39.

2. Caleb Williams Saleeby, *Sunlight and Health* (London: Nisbet & Co., 1923), 65.

3. J. H. Kellogg, *Light Therapeutics: A Practical Manual of Phototherapy for the Student and Practitioner* (1910; Battle Creek: Modern Medicine Publishing Company, 1927), 9.

4. J. H. Kellogg, *Heliotherapy, Phototherapy, and Thermotherapy*, 'Supplementary Chapters' to *A System of Physiologic Therapeutics*, ed. Solomon Solis Cohen. Volume IX, 'Hydrotherapy, Thermotherapy, Heliotherapy and Phototherapy' (London: Rebman, 1902), 221, 212.

5. C. W. Saleeby, in Rollier, op. cit., xxi, xvi.

6. Paul Fussell, *Abroad: British Literary Traveling Between the Wars* (New York: Oxford University Press, 1980), 15, 21.

7. Stephen Spender, *World Within World* (London: Hamish Hamilton, 1951), 107.

8. Adam Clapham and Robin Constable, *As Nature Intended* (London: Heinemann, 1982), 30.

9. 'Helios the Health Giver', *Health & Efficiency* (December 1927), 632.

10. Jan Gay, *On Going Naked* (London: Noel Douglas, 1933), 122.

11. Edward Carpenter, *Civilisation: Its Cause and its Cure* (1889), 44–5.

12. C. W. Saleeby, in Hans Suren, *Man and Sunlight* (Slough: Sollux, 1927), vii.

13. Heinrich Pudor, *Naked People: A Triumph Shout of the Future*, trans. Kenneth Romanes (Peterborough: Reason Books, 1998), 25.

14. Leonard Henslowe, 'Feel Pride of Body', *Health & Efficiency* (October 1931), 120.

15. Typescript rules held in the archive of British Naturism.

16. Hans Suren, op. cit., 1.

17. Arnold Lane, 'Prudery's Veil of the Sun', *Health & Efficiency* (October 1931), 81.

18. Frank Miles, 'Tan without tears: the scientific way to a sun-bronzed body', *Health & Efficiency*, 1940 summer special, 8.

19. John Fiske, 'Reading the Beach', in *Reading the Popular* (Boston: Unwin & Hayan, 1989), 56.

20. 'Bottled Sunlight', *Health & Efficiency* (May 1925), 202.

4. ESCAPE

1. Charles Baudelaire, 'Anywhere out of this world', in *The Poems in Prose*, trans. Frances Scarfe (London: Anvil, 1989), 191.
2. Edmund C. P. Hull, *The European In India, or Anglo-Indian's Vade Mecum* (London, 1871), 61.
3. Lady [Mary Elizabeth] Herbert, *Impressions from Spain* (London, 1867), 1, 223.
4. Alain Corbin, *The Lure of the Sea: The Discovery of the Seaside in the Western World 1750–1840* (Cambridge: Polity, 1994), 149, 152.
5. Thomas More Madden, *On Change of Climate* (London, 1874, third edition), 32.
6. John Richard Green, *Stray Sketches from England and Italy* (London: Macmillan, 1876), 31.
7. Sir James Clark, *The Sanative Influence of Climate: With an account of the best places of resort for invalids in England and the South of Europe* (1821; London: John Murray, 1841, third edition), 106.
8. Robert Louis Stevenson, 'The Stimulation of the Alps', in *The Works of R. L. Stevenson: Further Memoirs* (London: Heinemann, 1924), 155.
9. Dr James Henry Bennet, *Winter and Spring on the Shores of the Mediterranean* (London: J. & A. Churchill, 1875, fifth edition), 167.
10. Jost Krippendorf, *The Holiday Makers: Understanding the Impact of Leisure and Travel* (London: Heinemann, 1987), 29.
11. E. M. Forster, *A Room With a View* (1908; Harmondsworth: Penguin, 1986), 81.
12. Yvonne Cloud (ed.), *Beside the Seaside* (London: Stanley Nott, 1934), 17.
13. Green, op. cit., 74, 74, 72, 49.
14. Andre Gidé, *Oscar Wilde* (Paris, 1910), 30. Many thanks to Merlin Holland, Wilde's grandson, for pointing me to this passage.

15. André Gide, *L'Immoraliste*, trans. Dorothy Bussy (1902; Harmondsworth: Penguin, 1960), 25, 26, 35, 55, 51.

16. Sun-worship seems to have had a strong homosexual orientation in these pioneer days. Walt Whitman made naked sunbathing part of his nature cult; Edward Carpenter followed suit, seeing this as a means of curing civilization; and then Gide and Forster used it as a code for sexual liberation.

17. D. H. Lawrence, 'Sun', *The Tales of D. H. Lawrence* (London: William Heinemann, 1934), 740, 741, 747, 746, 744, 746, 755, 742.

18. F. Scott Fitzgerald, *Tender is the Night* (1934; London: Grey Walls Press, 1953), 83, 65, 63, 65.

19. *The Times*, 22 June 1923.

20. The American identity of the trendsetters was perhaps significant. America appears to have embraced sunbathing as a fad earlier than the UK, with fashionable characters in Fitzgerald's earlier novel *The Beautiful and the Damned* (1922) professing a desire to be deeply tanned; Kerry Segrave cites newspaper reports of sunbathing at Newport, Rhode Island, as early as 1908: Kerry Segrave, *Suntanning in Twentieth-century America* (Jefferson, NC: McFarland & Co., 2005), 9.

21. Quoted by Segrave, op. cit., 33, 37.

22. John K. Walton, *The British Seaside: Holidays and Resorts in the Twentieth Century* (Manchester: Manchester University Press, 2000), 66.

23. Laurie Lee, *As I Walked Out One Midsummer Morning* (Harmondsworth: Penguin, 1969), 136.

24. V. Bote Gomez and M. Thea Sinclair, 'Tourism: Supply and Demand', in *Tourism in Spain: Critical Issues*, ed. M. Barke et al. (Wallingford: CAB International, 1996), 66.

25. Figures quoted in M. Barke and L. A. France, 'The Costa del Sol', in Barke et al., *Tourism in Spain*, 289.

26. Robert Graves, *Majorca Observed* (London: Cassell, 1965), 7, 39.

27. Paul Theroux, *The Pillars of Hercules: A Grand Tour of the Mediterranean* (Harmondsworth: Penguin, 1995), 15.

28. Paul Fussell, *Abroad* (New York: Oxford University Press, 1980), 38.
29. Lawrence, op. cit., 744.

5. PLEASURE

1. Richard Hobday quotes Aretaeus of Cappadocia from the second century AD, who mentioned how 'lethargics are to be laid out in the light and exposed to the rays of the sun, for the disease is gloom': Richard Hobday, *The Light Revolution: Health, Architecture and the Sun* (Forres: Findhorn Press, 2006), 12.
2. David Hirshleifer and Tyler N. Shumway, 'Good Day Sunshine: Stock Returns and the Weather', *Journal of Finance*, 58/3 (2003), 1009–32; Sabrina Bruyneel et al., 'Why Consumers Buy Lottery Tickets when the sun goes down on them: The depleting nature of weather-induced bad moods', Erasmus Report Series Research in Management (September 2005).
3. Matthew C. Keller et al., 'A Warm Heart and a Clear Head: The contingent effects of weather on mood and cognition', *Psychological Science*, 16/9 (2005).
4. Daniel Nettle, *Happiness: The Science Behind Your Smile* (Oxford: Oxford University Press, 2005), 39.
5. Norbert Schwarz et al., *Well-Being: The Foundations of Hedonic Psychology* (New York: Russell Sage Foundation, 1999).
6. Needless to say, many of the generalizations I make about sunshine and happiness would not apply to countries where the sun beats down every day, and is an object of respect or even fear. You don't see many African Bushmen deliberately seeking the sun.
7. Matthew Collins, *Altered State: The Story of Ecstasy Culture and Acid House* (London: Serpent's Tail, 1997), 61, 280.
8. Ralph Moore from *Mixmag* magazine kindly supplied his all-time Ibiza sunshine top ten: (1) 'Everybody Loves the Sunshine'

(Roy Ayers); (2) 'The Sun Rising' (The Beloved); (3) 'Sun Is Shining' (Bob Marley); (4) 'Blister In the Sun' (Violent Femmes); (5) 'You Are the Sunshine of My Life' (Stevie Wonder); (6) 'Here Comes the Sun' (The Beatles/Nina Simone); (7) 'Sunshine' (Jay-Z); (8) 'Sunshine of Your Love' (Cream); (9) 'I Am the Black Gold of the Sun' (4 Hero); (10) 'The Sun Can't Compare' (Larry Heard).

9. Norman E. Rosenthal, *Winter Blues: Everything you need to Know to Beat Seasonal Affective Disorder* (New York: Guildford Press, 2006, revised edition), 64.

10. Hobday, op. cit., 29.

11. P. R. Mills et al., 'The effect of high correlated colour temperature office lighting on employee wellbeing and work performance', *Journal of Circadian Rhythms* (2007), 5.

12. D. Orentreich et al., 'Sunscreens: Practical Applications', in *Sun Protection and Man*, ed. Paolo U. Giacomoni (Amsterdam: Elsevier, 2001), 550; also Giaomoni, introduction to the same volume, viii.

13. 'Sunlight, Tanning Booths and Vitamin D', *Journal of the American Academy of Dermatology* (2005), 52, 868–76, 871; http://www.bma.org.uk/ap.nsf/Content/sunbma

14. S. R. Feldman, A. Liguori et al., 'Ultraviolet exposure is a reinforcing stimulus in frequent indoor tanners', *Journal of the American Academy of Dermatology* (2004), 51, 45–51, 46.

15. Paul C. Levins et al., 'Plasma beta-endorphin and beta-liptotropin response to ultraviolet radiation', *Lancet*, 2 (16 July 1983), 166; F. Greister et al., 'The effect of artificial light and natural sunshine upon some psychosomatic parameters of the human organism', in C. Helene et al., *Trends in Photobiology* (New York: Plenum Press, 1982), 465–84; M. Wintzen, B. A. Gilcrest, 'Proopliomelanocortin Gene in keratinocytes', *Journal of Investigative Dermatology* (1996), 106, 673–8; a study which questions the connection between UV and opioids is T. Gambrichler et al., 'Plasma levels of opioid peptides after sunbed exposures', *British Journal of Dermatology*, 147 (2002), 1207–11.

16. Feldman, Liguori et al., op. cit., 50.
17. Mandeep Kaur et al., 'Induction of withdrawal-like symptoms in a small randomized, controlled trial of opioid blockade in frequent tanners', *Journal of the American Academy of Dermatology*, 54 (April 2006), 709–11, 709, 710.
18. Mandeep Kaur et al., 'Side effects of naltrexone observed in frequent tanners: Could frequent tanners have ultraviolet-induced high opioid levels?', *Journal of the American Academy of Dermatology*, 52 (May 2005), 916.
19. What of the depleted ozone layer? According to Richard Hobday, 'there is certainly no evidence that the increase in melanoma rates in recent years is linked to the "hole" in the ozone layer': Hobday, op. cit., 59.
20. Jonathan Balcombe, *The Pleasurable Kingdom: Animals and the Nature of Feeling* (New York: Macmillan, 2006), and in an e-mail exchange.
21. John Hawk and Jane McGregor, *The Family Doctor Guide to Skin and Sunlight* (London: Dorling Kindersley, 2000), 14.
22. Jean Jacques Bonerandi, 'Sociological Perspectives on Suntanning', in Jean Paul Ortonne and Robert Ballotti (eds), *Mechanisms of Suntanning* (London: Martin Dunitz, 2002), 375.
23. Michael F. Holick, *The UV Advantage* (New York: Ibooks, 2003), 75, 13–14.
24. The US Sun Safety Alliance is a 'non-profit' organization 'dedicated to the task of reducing the incidence of skin cancer in America by motivating people to actively adopt and practice safe sun protection'. In 2005 it reported: 'Over the past year, the number of Americans using sunscreen, a primary protector against skin cancer, has declined. At the same time, more Americans, including children, are being diagnosed with skin cancer than ever before.' The major industry financiers of the association are Coppertone Skincare (whose very name indicates a wholly different attitude to tanning) and the National Association of Chain Drug Stores. Both have strong representation on the Alliance's board.

25. Oliver Gillie, *Sunlight Robbery* (The Health Research Forum, 2004), 1.
26. Andrew Ness et al., 'Are we really dying for a tan?', *British Medical Journal*, 319 (10 July 1999), 114–16; http://www.bmj.com/cgi/content/full/319/7202/114
27. John Gage, *Colour and Culture: Practice and Meaning from Antiquity to Abstraction* (London: Thames & Hudson, 1993); Victoria Finlay, *Colour: Travels Through the Paintbox* (London: Sceptre, 2002).

6. LOVE

1. John Ruskin, 'Of the Pathetic Fallacy', in Dinah Birch (ed.), *John Ruskin, Selected Writings* (Oxford: Oxford University Press, 2004), 71, 73.
2. John Ruskin, *The Storm Cloud of the Nineteenth Century* (Sunny Side, Orpington: George Allen, 1884), 13, 46.
3. When Bob Dylan was invited to host a radio show in 2006, he chose as the theme for his maiden appearance as a DJ 'The Weather' in songs.
4. John Milton, *Paradise Lost*, in John Carey and Alastair Fowler (eds), *The Poems of John Milton* (London: Longman, 1968), 961.
5. Nick Cave, 'The Secret Life of the Love Song', in *The Complete Song Lyrics, 1978–2001* (London: Penguin, 2001), 8, 13.
6. Nick Hornby, *31 Songs* (London: Penguin, 2003), 216.
7. Sir Philip Sidney, in Gerald Bullett (ed.), *Silver Poets of the Sixteenth Century* (London: Everyman Paperbacks, 1984), 206, 209.
8. William Shakespeare, *The Complete Works*, ed. Stanley Wells and Gary Taylor (Oxford: Oxford University Press, 1988).
9. Gavin Pretor-Pinney, *The Cloudspotter's Guide* (London: Sceptre, 2006), 60.
10. Jeff Green, *The Green Book of Songs by Subject: The Thematic Guide to Popular Music* (Nashville: Professional Desk Ref., 2002, fifth edition).

11. Jon Savage, *Teenage: The Creation of Youth, 1875–1945* (London: Chatto & Windus, 2007), argues for a much longer gestation period for the concept of the teenager than is usually acknowledged.

12. Roger Lax and Frederick Smith, *The Great Songs Thesaurus* (New York: Oxford University Press, 1989, second edition).

13. Whilst the Beach Boys were not entirely responsible for the new summer idyll, they certainly have claims to being its undisputed owners and propagandists. They were seriously fixated with the idea of summer, and Brian Wilson appears to be obsessed with the sun, as witnessed by their backlist titles: 'All Summer Long'; 'Keep an Eye on Summer'; 'Keepin' the Summer Alive'; 'Summer Means New Love'; 'Sunshine'; 'Your Summer Dream'.

14. Nik Cohn, *Awopbopaloobopalopbam-boom: Pop From the Beginning* (London: Pimlico, 1969), 100.

7. MEMORY

1. Ovid, *The Metamorphoses,* trans. Arthur Golding, ed. Madeleine Forey (Harmondsworth: Penguin, 2002), 34.

2. Evelyn Waugh, *Brideshead Revisited* (Harmondsworth: Penguin, 1982), 29, 32–3.

3. Bill Bryson, *Notes From a Small Island* (London: Black Swan, 1996), 29.

4. George Orwell, *Coming Up for Air* (Harmondsworth: Penguin, 1990), 37, 109.

5. Paul Fussell, *The Great War and Modern Memory* (1975; Oxford: Oxford University Press, 2000), 236.

6. L. P. Hartley, *The Go-Between* (Harmondsworth: Penguin, 1997), 87.

7. J. L. Carr, *A Month in the Country* (Harmondsworth: Penguin, 2000), 65.

8. As we saw earlier, the sun was newly valued in the interwar years, and a string of good summers in the 20s provided

occasion to express the new solar awareness. This pronounced solar exposure has helped develop, perhaps overdevelop, our picture of this epoch.

9. Philip Eden, *The Daily Telegraph Book of the Weather* (London: Continuum, 2003), vii–viii.

10. William Wordsworth, 'Ode: Intimations of Immortality From Recollections of Early Childhood', in Thomas Hutchinson (ed.), *Wordsworth: Poetical Works* (Oxford: Oxford University Press, 1969), 460.

11. Kenneth Grahame, *The Penguin Kenneth Grahame* (Harmondsworth: Penguin, 1983), 5.

12. Arthur Ransome, *Swallows and Amazons* (London: Jonathan Cape, 2004), 246.

13. Vladimir Nabokov, in James McConkey (ed.), *The Anatomy of Memory* (New York: Oxford University Press, 1996), 331.

14. Plato, *The Republic*, Book VII, trans. Desmond Lee (London: Penguin, 2003), 244.

15. Jacquetta Hawkes, *Man and the Sun* (London: Cressett Press, 1962), 40.

16. Henry Fox Talbot, *The Pencil of Nature* (London: Longman, Brown, Green & Longmans, 1844), no page numbers.

17. Henry Fox Talbot, 'Some Account of the Art of Photogenic Drawing, Or, The Process by which Natural Objects may be made to Delineate Themselves without the Aid of the Artist's Pencil', 31 January 1839, reproduced in Beaumont Newhall (ed.), *Photography: Essays and Images* (London: Secker & Warburg, 1980), 24.

18. Henry Fox Talbot, quoted in Gail Buckland, *Fox Talbot and the Invention of Photography* (London: Scolar Press, 1980), 76.

19. Oliver Wendell Holmes, in Newhall (ed.), op. cit., 73.

20. *Metaphors of Memory* is the title of a fascinating book by Douwe Draaisma, about the way different technologies, including photography, have been used to describe the way memory functions. *Metaphors of Memory: A History of Ideas about the Mind* (Cambridge: Cambridge University Press, 2000).

21. Marcel Proust, *Against Sainte-Beuve*, trans. John Sturrock (Harmondsworth: Penguin, 1988), 6.
22. Marcel Proust, *Remembrance of Things Past*, vol. 3, trans. C. K. Scott Moncrieff (Harmondsworth: Penguin, 1983), 716.
23. Trevor A. Harley, '"Nice Weather for the Time of Year": The British Obsession with the Weather', in Sarah Strauss and Benjamin S. Orlove (eds), *Weather, Climate, Culture* (Oxford: Berg Publishing, 2003), 113.
24. Bryson, op. cit., 124, 124–5. John Walton remarks how Bryson's reading matter portrays 'a seaside world which was already endangered, if not on nostalgia-fed life support systems, by the later twentieth century . . . This is the innocent and timeless vision of the seaside which the post-war generation embroidered from its own memories of childhood holidays, at resorts which were just beginning to pass into the time-warp from which it was to prove difficult to rescue them': John K. Walton, *The British Seaside: Holidays and Resorts in the Twentieth Century* (Manchester: Manchester University Press, 2000), 3.
25. An excellent book on the 'Great British Seaside Holiday' is Steven Braggs and Diane Harris, *Sun, Sea and Sand* (Stroud: Tempus, 2006). They have a chapter on how sun-worship impacted on the marketing of resorts. The book is a great source of holiday nostalgia.
26. 'Don't mention the weather' appears to be the rule in marketing Britain these days. The Enjoy England website, which promotes England as a holiday destination to the domestic market, makes no mention of it.
27. The co-writer of Coppola's classic, John Milius, also wrote and directed *Big Wednesday* (1978), the only other serious surfing movie to gain mainstream popularity and critical acclaim. It is a film about dodging the draft as well as catching the waves.
28. Trevor Harley is very good on how our image of a 'traditional' Christmas is based on Dickens's own selective memory of a few unrepresentative snowfalls: 'Nice Weather', in Strauss and Orlove (eds), op. cit., 109–11.

29. Philip Eden, *Change in the Weather: Weather Extremes and the British Climate* (London: Continuum, 2005), 65.

30. John Berger, *Ways of Seeing* (Harmondsworth: Penguin, 1972), 139, 131.

8. LIFE

1. Robert Gittings (ed.), *Letters of John Keats, A Selection* (Oxford: Oxford University Press, 1970), 87.

2. Jonathan Bate, *The Song of the Earth* (London: Picador, 2001), 120.

3. Peter A. Stott et al., 'Human contribution to the European heatwave of 2003', *Nature*, 432 (2 December 2004), 610–14, 613.

4. Robert Henson, *The Rough Guide to Climate Change* (London: Rough Guide, 2006), 115.

5. Kate Fox, *Watching the English: The Hidden Rules of English Behaviour* (London: Hodder, 2004), 304, 32, 392.

6. *Daily Express*, Monday 16 April 2007. The temperature didn't reach anywhere near this again until August and then for one feeble day.

7. Richard Reeves, 'The Politics of Happiness: a Discussion Paper' (New Economics Foundation, 2003).

8. www.britishweatherservices.co.uk

9. 'Forget sun, sea and sand. This year business is more soup by the telly': *Guardian*, 26 July 2007. http://business.guardian.co.uk/retail/story/0,,2134982,00.html#article

10. Press release issued by the IPP. www.ippr.org and data is available from www.bbc.co.uk/bornabroad

11. Henry Buller and Keith Hoggart, *International Counterurbanization: British Migrants in Rural France* (Aldershot: Avebury, 1994), 1, 4.

12. Giles Tremlett, *The Ghosts of Spain: Travels Through a Country's Hidden Past* (London: Faber & Faber, 2006), 109, 108.

13. George Orwell, 'Pleasure Spots', originally in *Fortune*, January 1946, http://www.orwell.ru/library/articles/spots/english/e_spots

14. Tremlett, op. cit., 123.

15. J. G. Ballard, *Cocaine Nights* (London: Flamingo, 1997), 180, 218.

16. In Buller and Hoggart's post-1988 survey, 39.1 per cent mentioned property price and 22.7 per cent climate as their main motivators for moving to France, showing that this is in line with overall trends, but not so very different from those who move to southern Spain. The majority of people indicated that 'their main reasons for migrating to France were the need for a change in lifestyle, a better climate, a cheap place to live or a more relaxed mode of living': Buller and Hoggart, op. cit., 86. The Costa del Sol equivalent of this study is Karen O'Reilly, *The British on the Costa del Sol* (London: Routledge, 2000), providing an anthropological study of a community in Fuengirola, where she lived for a number of years.

17. Peter Mayle, *A Year in Provence* (London: Hamish Hamilton, 1989), 2, 80.

18. Peter Mayle, *Toujours Provence* (London: Penguin, 2001), 151.

19. Mayle, *Year in Provence*, 136.

20. George Bernard Shaw, Preface to *Pygmalion* (1916).

21. Buller and Hoggart, op. cit., 118.

22. Van Gogh, *c.* 16 June 1888, *c.* 18 June 1888, cited in Robert Pickering, *Van Gogh in Arles* (New York: Metropolitan Museum of Art, 1984), 22, 27.

23. Letter to Bernard, *c.* 18 August 1888, in *Vincent Van Gogh: Letters from Provence*, selected and ed. Martin Bailey (London: Collis & Brown, 1990), 58.

9. RETURN

1. 'A Global Projection of Subjective Well-being: A Challenge to Positive Psychology?', *Psychtalk*, 56 (2007), 17–20. I wrote to Adrian White, the author of this study, at the University of Leicester, asking whether weather had been taken into consideration in his study, and he very kindly offered to factor it in

retrospectively. On the basis of the average yearly sunshine figures for most of the countries in his rankings, he performed a number of calculations. This is what he found: 'Sunshine levels are negative associated with international levels of subjective wellbeing (happiness)', as a glance at the geography suggests. Factors like health, poverty and education predicted 48 per cent of the variance in happiness. 'However, when we control for the effect of poverty (remove the effects of poverty on happiness levels – using GDP per capita) we actually find that the top ten sunny countries are 60 per cent happier than the bottom ten sunny countries. Poverty is still the hugely significant predictor, but when you remove the effects of this, sunshine has a significant effect.' Poor but sun-blessed. I suppose that's how mankind started out.

2. 'Does living in California make people happy? A focusing illusion in judgments of life satisfaction', David A. Schkade and Daniel Kahneman, *Psychological Science*, 9/5 (September 1998), 340–46.

3. Philosophy and physiology are in agreement on this issue. Daniel Nettle explains how wanting and liking are controlled by different chemicals in the body, and that addictive substances such as tobacco, cocaine and heroin drive people to pursue something they do not always derive pleasure from. If sunshine is addictive, through the stimulation of opioids in the body, then it would be subject to this mechanism. Serotonin is not addictive, and therefore those who live under sunny skies take their sunlit environment for granted, and are less inclined to crave direct UV stimulation: Daniel Nettle, *Happiness: The Science Behind Your Smile* (Oxford: Oxford University Press, 2005), 126–34.

4. 'Science Tracks the Goodlife', *San Francisco Chronicle*, 24 December 2000. The article was reporting a survey conducted for the paper by a psychologist reviewing decades of reports on national indexes of happiness: http://sfgate.com/cgi-bin/article.cgi?file=/chronicle/archive/2000/12/24/MN165379.DTL

5. Lucretius, *On the Nature of Things*, Book II, trans. H. A. J. Munro (Cambridge: Deighton Bell & Co., 1886), 28–9.

6. John Keats, in Robert Gittings (ed.), *Letters of John Keats, A Selection* (Oxford: Oxford University Press, 1970), 283.

7. According to Bate, 'the good summer and clear autumn of 1819 very literally gave him a new lease of life': Jonathan Bate, *The Song of the Earth* (London: Picador, 2001), 105.

8. John Barnard (ed.), *John Keats: The Complete Poems* (Harmondsworth: Penguin, 1973), 434.

9. Louis MacNeice, 'The Sunlight on the Garden', *The Collected Poems of Louis MacNeice* (London: Faber & Faber, 1979). Coincidentally, MacNeice wrote those lines when he was living in a garden flat in what is now called Keats Grove, in Hampstead, fifty yards from where Keats had spent his last years in England, and where he wrote his 'Ode to a Nightingale', another poetic monument to mutability.

10. John Milton, *Paradise Lost*, in John Carey and Alastair Fowler (eds), *The Poems of John Milton* (London: Longman, 1968), 562.

11. Ben Jonson, *The Alchemist*, II.1.46–51, ed. F. H. Mares (Manchester: Manchester University Press, 1979), 47.

ACKNOWLEDGEMENTS

I'd like to thank all those who provided valuable information, advice, time or support in helping me write this book: Daniel Nettle, Michael Farrar, Richard Hobday, Steve Hayes, John Eagles, Oliver Gillie, Ciaran Mooney, Angela Wright, Peter Stott, Ken Pritchard, Matt Havercroft and Amanda Lamb, Kevin Bacon, Robert Altman, Andrew Welch, Frances Baines, Jonathan Balcombe, Philip Eden, Trevor Harley, Frank Somers, Peter Mills, Jim Dale, Liz Settle, Harriet Quick, Anna-Marie Solowij, Peter Borsay, Ralph Moore, Jennifer Ballatine Perera, Ros Lavine, Jeremy Hildreth, Belinda Hubball, Matt Silcock and Jeremy Shaw. Thanks to Dan Hind, Lisa Tennant, Ben Hervey, Sarah Adam, Sonia Massai, Richard McCabe, Tommy and Holly Crocker, Sean Kingsley, Merlin Holland, Kate Garvey, Sam Jordan, Sarah Hocombe; to all at Radio Europe Mediterraneo, and the good people of Marbella.

Special thanks go to Laura Barber, Clara Farmer, Simon Bradley, Astrid Pyrko, Corinne Chabert, Susie Cornfield, Laura McCarthy, and Cecilia Mackay for their kindnesses; to Helen and Phil in France and to Danny King in Madrid for providing shelter from the English climate; to John Everblest of the Lifestyle Laboratory, for allowing me to use his images from the Kellogg archive, and to Dr Adrian White for his help crunching figures on the global map of happiness. Big thanks to Patrick Walsh, my agent, who toiled to ensure my proposal was fighting fit; to all at John Murray for turning it into a book; and to Lloyd Northover, especially Jim and Jim, for being flexible and understanding about my 'day job' over the last year.

Big thanks to friends and neighbours for putting up with my monomania over the years: Nickie and JJ, Douglas and Georgia, Paula and Fabrizio, Emma and Mikey, and all the sun-worshippers at the Mills. Love to my parents, and to Rowan for everything.

PHOTO CREDITS

(p. 27) *The Graduate*: courtesy Embassy/The Kobal Collection; (p. 28) *Easy Rider*: courtesy Columbia Pictures/The Ronald Grant Archive; (p. 29) *Withnail and I*: courtesy HandMade Films/The Ronald Grant Archive; (p. 37) Edwardian Ladies in Brighton, courtesy of the Royal Pavilion & Museums, Brighton and Hove; John Harvey Kellogg Archive images are by courtesy of John Everblest at www.lifestylelaboratory.com; (p. 93) Tom Purvis poster of sunbathers for LNER, courtesy of the National Railway Museum/Science and Society Picture Library; (p. 98) Happy holidaymakers at Brighton in the 1930s, courtesy of the Royal Pavilion & Museums, Brighton and Hove; (p. 115) Mabli's sun, courtesy of Mabli; (p. 117) 'Dance' (hippies in Golden Gate Park) © and by courtesy of Robert Altman; (p. 123) 'Sunbathers at Lake Stratzersee', courtesy of National Media Museum/Science and Society Picture Library; (p. 129) *Physignathus lesueurii* (Basking Easter Water Dragon) © Frances M. Baines; (p. 141) *The weather project* 2003, Turbine Hall Tate Modern London (the Unilever Series), monofre-

quency light, reflective panel, hazer, mirrored foil, steel: ©
Olafur Eliasson, photograph © Philippe Bryse; (p. 166) *Beach
Party*, courtesy of the Kobal Collection; (p. 181) 'Redcar', LNER
poster, courtesy of the National Railway Museum/Science and
Society Picture Library; (p. 197) Apollo the Sun God, courtesy
of Eastbourne Visitor Centre and the Royal Pavilion &
Museums, Brighton and Hove; (p. 200) *The Endless Summer*,
© and courtesy of Bruce Brown Films Inc; (p. 203) Brighton
Bank Holiday girls in the rain, courtesy of Royal Pavilion &
Museums, Brighton and Hove; (p. 213) 'Sunbather on Clapham
Junction Station' © and courtesy of Chris Mole; (p. 236)
Vincent Van Gogh, *View of Les Saintes-Maries*, 1888; Oskar
Reinhart Collection, Winterthur, courtesy of akg-images.

I'd like to acknowledge permission to reproduce copyright
material from the following:

Quotations from the *Daily Mail* are by kind permission of the
newspaper. Quotations from the *London Evening Standard*
are by kind permission of the newspaper. Lines from 'Here
Comes Summer' by Jerry Keller are quoted by kind permission of Lynn Hatch Music/International Music Network.
Lines from 'The Sunlight on the Garden' by Louis MacNeice
are from *Louis MacNeice: Collected Poems* published by Faber
& Faber, and are quoted by kind permission of David Higham
Associates. Thank you to Macmillan Publishers Ltd for allowing me to quote from the novelisation of *Sexy Beast*, by
Andrew Donkin (Macmillan, 2000).

Every effort has been made to clear permissions. If permission has not been granted please contact the publisher who
will include a credit in subsequent printings and editions.

INDEX

INDEX

Index